INDUSTRIAL-ARTS ELECTRICITY
BASIC ELECTRONICS

COURTESY, GENERAL ELECTRIC

Hoover Dam and Lake Mead from the Air. This 726.4-foot-high dam in Black Canyon of Colorado River has formed a lake 115 miles long.

Fifth Edition

Industrial-Arts
ELECTRICITY

Basic Electronics

By **CLIFFORD K. LUSH**

Former Director, Vocational Education and Industrial Arts, Minneapolis Public Schools, Minneapolis, Minn.

and **GLENN E. ENGLE**

Former Instructor in Electricity, Minneapolis Public Schools, Minneapolis, Minn.

Chas. A. Bennett Co., Inc. *PUBLISHERS*

PEORIA, ILLINOIS

Copyright, 1971

Previous copyrights 1946, 1956, 1959, 1965

BY CHAS. A. BENNETT CO., INC.

Use of all problems, jobs, diagrams, photographs, and other informative or practical shop material in this book is specifically restricted under the Copyright Law of the United States, except for direct classroom or similar learning situations; and reproduction in any form, including simple types of duplication, is expressly forbidden unless written permission is secured from the publishers.

147K110

Library of Congress Catalog No. 65–18201

SBN 87002-103-6

Fifth Edition

PRINTED IN THE UNITED STATES OF AMERICA *in Bodoni, Modern Lino No. 21, and Futura types.*

CONTENTS

Acknowledgments   7

Suggestions to the Instructor   9

CHAPTER I

Introduction to the World of Electricity 11
    Electricity is everywhere — uses to which it may be put — occupations to which it applies — how you learn about electricity the "natural way."

CHAPTER II

Magnetism and How It Is Related to Electricity 19
    Many uses in everyday life — kinds of magnets — how the force of magnetism operates — nature of magnetized material — earth as a magnet — inducing magnetism.

CHAPTER III

Sources of Electricity and Electrical Energy 28
    Two types of electricity — static electricity — current electricity — direct and alternating current — generators — electricity in cells — series and parallel wiring — voltage and amperage — symbols.

CHAPTER IV

Electromagnetics or Magnetism Induced by Electrical Flow 37
    How the electromagnet is formed — the solenoid or coil with movable core — induction coil — electrical "shock" — induction coil, voltage, spark coil, and transformer.

CHAPTER V

The Flow of Electricity and Conducting Materials 47
    Conductors and insulators — electron theory, kinds of wire, insulation, splicing — symbols for wiring — switches — electric bell and buzzer — terminals — using dry cell or transformer.

CHAPTER VI

Low-Voltage Circuit Wiring of Signal Devices 57
    Familiar types of circuits — bell-wiring problems — simple signal alarms and controls.

CHAPTER VII

Heat From Electricity Applied to Everyday Problems 66
    Meaning of "resistance" — resistance measured in Ohms — resistance and voltage — the rheostat — Ohm's law — heating appliances — fuses — other applications of heat from electricity.

CHAPTER VIII

Lighting With Electricity and Electrical Lighting Devices 78
    The electric-light bulb — portable lighting — low-voltage lighting circuits — two-way switch for lights — lights in parallel and series — meter reading — extension cords — sockets — feed-through switches and other problems.

CHAPTER IX

House Wiring: Electrical Conduits and Switches 91
    Kinds of wire used — joints — soldering — house fixture wiring — switches — conduit wiring — bending conduit — pulling wires through conduits.

CHAPTER X

Communication by Means of Electrical Transmission 113
    Blinker lights — the telegraph, and telegraph codes — the telephone — radio — use of electricity for other means of communication — television.

CHAPTER XI

Electrical Power and the Electrical Generator and Motor 129
    Horsepower — the generator — the electric motor — electricity in the automobile — the storage battery — glow plug engine — railroads and electricity — electronics — atomic power — lasers.

Examinations 155
    Test of reading ability in electricity — final comprehensive examination (example of).

Index 159

ACKNOWLEDGMENTS

The writers wish to express their appreciation to the manufacturers who furnished pictures for many of the illustrations and to Lee Moren, teacher of electricity at Folwell Junior High School, Minneapolis, Minn., for his assistance with the content on electronics.

Western Electric Co., Minneapolis, (telephone)
Grant Storage Battery Co., Minneapolis, Minn.
Westinghouse, Pittsburgh, Penna.
A. G. Redmond Co., Owosso, Mich. (electric motor)
D. Onan and Son, Minneapolis, Minn. (gas electric generators)

Automatic Electric Co., Chicago, Ill. (automatic switch)
Northern States Power Co., Minneapolis, Minn.
M. B. Austin Co., Chicago, Ill. (conduit)
Square D Co., Chicago, Ill.
Dings Magnetic Separator Co., Milwaukee, Wis.
Cutler Hammer, Inc., Milwaukee, Wis.
Minneapolis Electric Machinery Mfg. Co., Minneapolis, Minn.
General Electric, Schenectady, N.Y.
Illinois Electric Porcelain, Macomb, Ill.
Sterns Magnetic Mfg. Co., Milwaukee, Wis.
Minnesota Mining and Manufacturing Co., St. Paul, Minn.
Honeywell, Minneapolis, Minn.
Bell Telephone Systems

List of Color Illustrations

Electric power from water and steam, 8A
Worker in atomic power plant, 8B
Electricity from atomic power, 8C
Control room in atomic power plant, 8D
Computer room in atomic power plant, 8D
Repairing power line, 8E
Assembling color-TV sets, 8E
Checking light sensors, 8F
Cooling and heat control center, 8F
Installing a thermostat, 16A
A motor-generator unit, 16A
Model airplane remote control contest, 16B

Rain or shine the contest is held, 16C
Project in electronics course, 16D
Communications laboratory, 32A
Automotive testing equipment, 32B
Circuits in a television receiver, 112A
Tracing a circuit, 112A
TV troubleshooting, 112A
Soldering a defective circuit, 112B
Checking with an oscilloscope, 112B
TV repairing, 112B
Underseas cable, 128A
Laying an underground cable, 128A
Computer tape of stored information, 128A
Emergency telephone repair crews, 128B
Telephone microwave tower, 128B

Electric power from water and steam on the upper Mississippi.

Technician working in atomic power plant.

COURTESY NORTHERN STATES POWER CO.

Electricity from atomic power.

The control room in an atomic power plant.

COURTESY NORTHERN STATES POWER CO.

Computer room in an atomic power plant.

COURTESY NORTHERN STATES POWER CO.

COURTESY NORTHERN STATES POWER CO.

Workmen repairing power line.

Electronics workers on a line, assembling color-TV sets.

COURTESY, LEE MOREN

COURTESY, HONEYWELL

Control center for the cooling and heating system of a large building complex. This technician works full time interpreting the data electronically supplied to this center.

An electronics technician checking light sensors. These tubes "see" gas flames and serve to shut down furnaces in emergencies.

COURTESY, HONEYWELL

CHART A
Engineers and Technicians

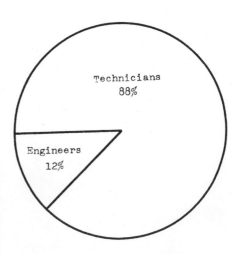

Technicians 88%

Engineers 12%

Number of professionally trained electrical engineers compared to trade-technical trained workers.

SUGGESTIONS to the Instructor

Since the electrical industries employ so many people, teachers of beginning electricity courses may wish to provide their students with information on electrical occupations. In addition, teachers may want to emphasize the relative number of workers in these occupations. This first course will certainly also include information and training that will help the students live more knowledgeably and effectively in a world of things electrical.

Of the some fifty million employed civilian men in the United States, age sixteen and over, it is estimated that two million work in the electrical industry. Very few students in the average high school electricity class, however, will later become professional electrical engineers or scientists. A much larger number will become technicians and tradesmen, who have less than four-year degrees. Chart A compares the number of technicians and engineers in the industry.

Major Occupations on the Technical Trade Level

There are about one and a half million workers in the electrical industry who do more than assembly work but are below the professional level. This figure does not include workers, such as appliance repairmen, x-ray technicians, and general auto mechanics, who need only a little knowledge of electricity. Within this technical field, there are eight occupational families. These are shown in Chart B, which was compiled from the U. S. Dept. of Labor Statistics *Occupational Outlook Handbook*. All

Chart Area	Percent of Total	Type of Employment
A	25%	Service maintenance workers in factory, home, and office (300,000)
B	17%	Telephone industry workers (200,000)
C	14%	Electricians in the construction industries (160,000)
D	11%	Servicemen of radios, TV's, etc. (130,000)
E	9%	Technicians in electronic manufacturing (100,000)
F	9%	Technicians in general electrical manufacturing (100,000)
G	8%	Armed services and space program technicians (95,000)
H	7%	Miscellaneous technicians, such as medical, aircraft, etc. (90,000)

CHART B
Electrical Technicians

of these require a broad basic course in elementary electricity.

On Chart B, note that areas A, B, and D represent technical service occupations. The remaining ones represent technicians in manufacturing, construction industries, and miscellaneous others. These two categories have about the same number of employees. (See Chart C.)

Those who are service technicians need a background in electricity, even though mechanics is their major concern. The number of men who are trained in electronics compared to those who have a minimum of electrical training is shown in Chart D.

The primary work of all servicemen is to troubleshoot, repair, replace, install, and adjust.

What About House Wiring?

Teaching students about wiring is quite important. This knowledge will enable them to use proper procedures when an electrical problem occurs.

Note that this text offers the information about electricity and tools that is needed by a career-minded student. Yet, it also offers training for the typical consumer of electricity.

For many students, this first course will be their only opportunity to learn about the many electrical occupations. It may also be their only opportunity to learn the safe and intelligent use of the electrical devices encountered every day of their lives.

The Workbook for Industrial-Arts Electricity

The workbook for this text will be valuable in this first course for two reasons. It will be useful for enrichment and will also allow the teacher more time for the individual student.

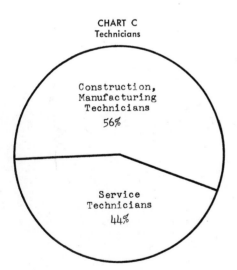

Number of service electrical technicians compared to construction and manufacturing technicians.

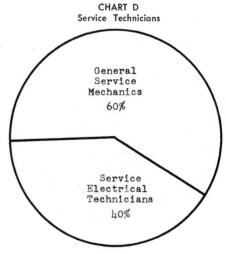

Number of general service mechanics with a minimum of electrical training compared to service electrical technicians, who are trained in electronics.

Chapter I

INTRODUCTION to the World of Electricity

Electricity is everywhere—but no one knows what it is! Uses to which it can be put—occupations to which it applies—how you learn about electricity the "natural way."

ELECTRICITY is everywhere. It is used in our homes and in buildings where we work or play or go to school, and as a means of transportation and communication. In fact, electricity is given off each time our heart beats and each time we move a muscle or each time we think! When you walk across the floor or rub one object against another, "electrons" from the rug or the object are disturbed or collected to unbalance the normal charge or electricity of the objects and we say that electricity is "produced," when we see sparks jump. The spark is the collection of "electrons" showing their

Factory workers building large generators.

COURTESY, WESTINGHOUSE

COURTESY, WESTINGHOUSE

Electrical tradesman at work on a large alternating current generator.

COURTESY, WESTINGHOUSE

Electrical tradesman installing a part of a giant switch to control 230,000 volts.

Mental abilities are just as important to success in electronics as hand skills. This young man is studying electronics. He is learning information and techniques which will prepare him for a career as a technician.

COURTESY, NORTHWESTERN ELECTRONICS INSTITUTE

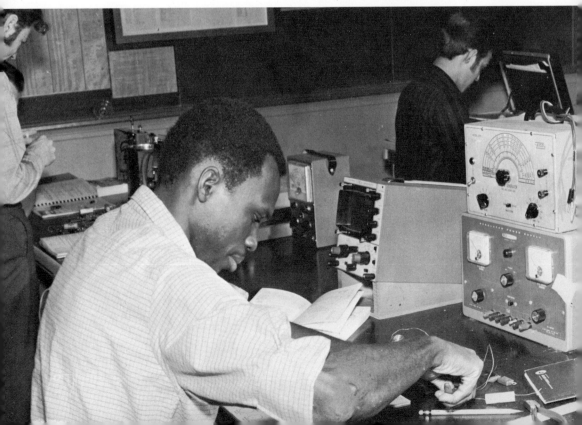

COURTESY, WESTINGHOUSE

Electrical engineers in conference.

haste to rush back into a "balanced state" again. But more about all this later.

Electricity has always been with man, but only within the last few decades has he learned how to use it. Today, the use of electricity is as commonplace as the air we breathe.

You are about to learn how man uses electricity in his daily life; how the tremendous power of electricity is "made" and controlled; how electrical machines and appliances found in the typical home operate, and how to care safely for them.

You will learn something about the fields of work in electricity at which men and women earn a living so that you may compare your interests and abilities with those required for success in such occupations. Should you ever decide that you would like to enter an occupation concerned with electricity, charts are included in this chapter which show the preparation needed. Most of the "roads" to the electrical fields of work lead through high school first and then to schools or colleges offering specialized courses. In high school, you should continue to explore still further your interests and abilities in electricity to see if they are genuine, lasting, or sufficient for success. Take mathematics and sciences and shopwork, because better-than-average success in these subjects

Electrical engineers at work.

COURTESY, WESTINGHOUSE

14

Electronic brain used in man-flight to stabilize the capsule. The technician is making a final check of all circuits.

COURTESY, HONEYWELL

Inspector making a final check on a missile guidance brain. The system supplies signals to the automatic pilot to guide a missile during its first few minutes of flight.

COURTESY, HONEYWELL

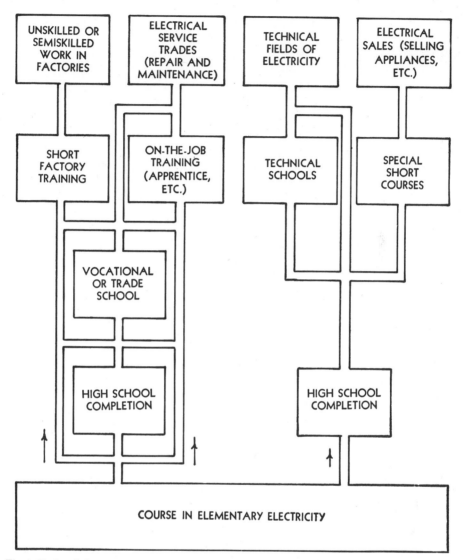

Fig. 1. Paths of electrical occupations requiring technical training rather than college training.

is necessary to be successful in all the electrical fields, with the possible exception of factory jobs which consist primarily of assembling or producing parts of electrical appliances.

WORK NOT REQUIRING COLLEGE EDUCATION

Trace the path of entry in Fig. 1 to the broad fields of electrical occupations which, for entrance, require an

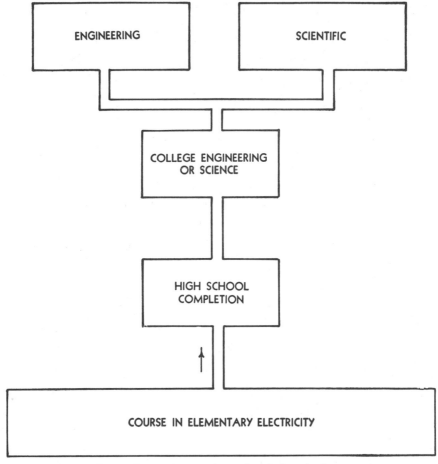

Fig. 2. Occupations, paths to electrical *professions*.

education of less than college grade.

The more training you have the greater are your chances for success and future promotion in the electrical fields.

Factory Work, Service Trades

The typical work in factories making electrical machines is routine assembling of parts. In the service trades, the work consists of installing or repairing or adjusting electrical machines. House wiring, radio and motor repair, and maintenance are examples of work in the service trades.

Technical Fields

In the technical fields, the highly specialized work consists of such jobs as operating an X-ray machine in a hospital, or repairing specialized sorting, cleaning, or conditioning equipment.

People who make specialized elec-

COURTESY DUNWOODY INSTITUTE, MINNEAPOLIS

Installing a thermostat.

Operating a motor-generator unit.

COURTESY MINNEAPOLIS VOCATIONAL HIGH SCHOOL

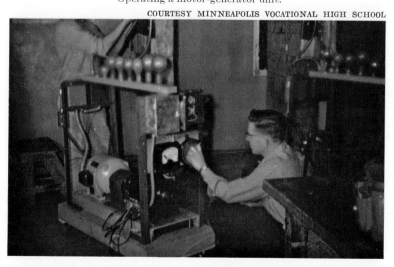

An entry in an annual model airplane remote control contest. Navy personnel are judges.

Rain or shine the contest is held.

NATIONAL MODEL AIRPLANE REMOTE CONTROL CONTEST

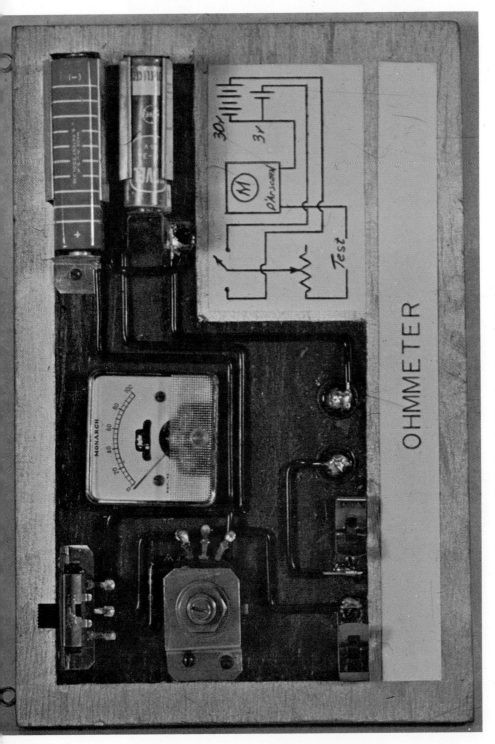

Student project in elementary electronics course.

INTRODUCTION TO THE WORLD OF ELECTRICITY

trical equipment need workers who understand their machines to sell their products.

WORK REQUIRING COLLEGE EDUCATION

The next chart in Fig. 2 shows the fields of work in electricity requiring a college education of from four to six years.

A student desiring to become an electrical engineer or scientist must have high marks in mathematics and science and possibly mechanical drawing to be successful.

Electrical engineers design the machines for the factories to build and the salesman to sell and the tradesman to install and maintain. The engineer makes practical use of theories and discoveries developed or made by the scientists. The electrical scientist works in laboratories, experimenting and testing and developing new ideas and principles about electrical action. Many engineering graduates enter the sales fields and technical fields, but very few go into the trades.

HOW YOU LEARN ABOUT ELECTRICITY

Learning about electricity is somewhat like building a block tower. To understand the material in one chapter it is necessary that you know what is in the previous chapter. Each chapter is built upon the chapter before it.

The jobs are included to help you understand the principles related to the job; to assist you in remembering the principles because we "learn by doing"; to teach you many hand skills; and to give you the opportunity to match the type of thinking required for success in several occupations in the electrical field with the type of mental equipment you may have.

Do you like to solve maze puzzles? If so, and you do them rapidly, it may indicate that you have the type of mentality to become an electrician.

INDUSTRIAL-ARTS ELECTRICITY will acquaint you with the major occupations of workers concerned with electricity and will teach you to understand and use intelligently the hundreds of things electrical which surround you in everyday life. The study of electronics is concerned with the flow of electrons. The flow of electrons is the flow of electricity through solids, gases, and even liquids, a new and exciting science.

Learning to become a radio repairman.   Technician checking a circuit.

COURTESY, WORK OPPORTUNITY CENTER, MINNEAPOLIS

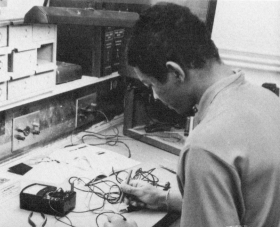

COURTESY, HONEYWELL

Fig. 3. A test engineer working on special equipment, checking the current balance transmitter. The transmitter controls the flow of current to all power units in a large factory.

How many of these are in your home?

Air Conditioners	Electric Bed Coverings	Humidifier	Sander
Attic Fans	Electric Clocks	Ice Cream Freezer	Sandwich Toaster
Other Fans	Electric Hobby Equipment	Incinerator	Sewing Machine
Auto Engine Heater	Electric Lawn Mower	Intercom System	Shaver
Automatic Sauce Pan	Electric Skillet	Ironer	Shears
Battery Charger	Electric Toys	Juicer	Shoe Shiner
Blender	Exhaust Fan	Knife Sharpener	Soldering Iron
Bottle Warmer	Flood Lights	Lawn Edger	Space Heaters
Can Opener	Food Chopper	Lawn Mower	Sterilizer Lamp
Christmas Lights	Food Mixer	Night Lamp	Sun Lamp
Clock-Radio	Food Warmer	Ozone Lamp	Tape Recorder
Clothes Dryer	Freezer	Paint Sprayer	Television
Clothes Washer	Furnace Motor	Phonograph	Timer
Coffee Maker	Garage Door Opener	Photo Flood Lamps	Toaster
Corn Popper	Germicidal Lamp	Portable Mixer	Typewriter
Deep Fryer	Grill	Power Tools (How many?)	Vacuum Cleaner
Defroster	Hair Dryer		Vaporizer
Dehumidifier	Hand Irons	Projector	Vibrator
Demother	Heating Pad	Radios	Waffle Maker
Door Bells or Chimes	Heat Lamp	Range	Water Heater
Dishwasher	Hedge Clipper	Record Player	Water Pump
Disposal Unit	Hot Dog Cooker	Refrigerator	Water Softener
Egg Cooker	Hot Plate	Roaster	Waxer-Polisher
Electric Barbecue Starter	House Number Sign	Rotisserie	

On a sheet of paper, *guess* the number of electrical appliances and conveniences (electric, except for lights) in your home. Take time to visualize as you make your list. Now check the items and on your paper write down those you *actually have* at home. Count each radio or TV, if you have more than one. How does your actual total compare with your first guess?

Chapter II

MAGNETISM and How It Is Related to Electricity

Many uses in everyday life—kinds of magnets—how the force of magnetism operates—nature of magnetized metal—earth as a magnet—inducing magnetism.

FEW people realize the uses made of magnets in everyday life. For example, many butchers use *magnetized steel sharpeners* to help draw and keep the knife on the sharpener. The upholsterer uses a *magnetized hammer* to hold his tacks. *Magnetic markers* show the positions of commercial planes in flight on a steel-backed map in the flight-control centers of large airports. Thousands of other uses of magnets might be given, but let us go on and get the full story of magnetism.

Thousands of years ago, as the story goes, shepherds used iron to tip their staffs. One shepherd, while roaming through the hills, noticed in one locality that the tip of his staff was covered with small hard particles. Being curious, he dug into the earth and found what he thought were rocks, which not only clung to the iron on his staff but also to the nails in his shoes. The ore he discovered has been named *magnetite* or *loadstone* — formally spelled lodestone — and one of the earliest practical uses made of the ore was in making compasses. For centuries, the ore was explained as a "magic force." Some people even thought its magic would cure a toothache or heal burns. Its use in the compass was known the world over by the time of Columbus. Loadstone or magnetite is classified as a *natural magnet*.

PERMANENT ARTIFICIAL MAGNETS

Artificial magnets may be made by stroking a piece of steel with a piece of loadstone; with this piece of steel, another piece may be magnetized, and so on. See Fig. 4. Soft iron (mild steel) can be magnetized very easily, but it will lose its magnetism as soon as the loadstone has been removed. A very hard steel is more difficult to magnetize than mild steel but, when a piece of hard steel has been magnetized, it is called a *permanent magnet*

INDUSTRIAL-ARTS ELECTRICITY

Fig. 4. Loadstone and a bar magnet.

Fig. 5. Varied shapes of permanent magnets.

because hard steel will hold a small amount of magnetism permanently. The steel may be in the form of a straight bar; or it may be bent much like a horseshoe or ring, with the ends or poles a short distance apart. Any magnet has a much stronger field when the ends, or poles, are close together, thus reducing the air gap. See Fig. 5.

Permanent magnets may be made by running an electric current through a coil of insulated wire wound around the piece of hard steel to be magnetized. See Fig. 6.

A piece of soft iron or steel placed across the ends of a horseshoe magnet helps the magnet to hold its magnetism when not in use. This piece of steel is called a *keeper*.

POLES

If a magnet is dipped into iron filings or into a box of tacks, most of the filings or tacks will cluster near the ends, with only a few near the middle. Magnetism, then, is strongest at the ends where it enters and leaves the magnet. These ends are called the *poles* of the magnet. See Fig. 7.

If the bar magnet is hung by a string so that it swings freely (Fig. 8), with no other metal or magnet near,

Fig. 6. Magnetic coil to magnetize a steel needle. First insert the needle, then make the connection. Connect for five seconds only.

Fig. 7. Iron filings on a bar magnet.

MAGNETISM

it will come to rest with one end pointing north and the other end pointing south. Such action is due to the influence of the earth's magnetism, which is everpresent on the earth's surface. Therefore one end of a bar magnet is usually marked "N" or the north-seeking pole, and the other is marked "S" or the south-seeking pole. A magnet is said to have *polarity,* the condition of having two opposite electrical or magnetic poles.

But the two poles of a magnet are not alike!

In fact, they are very different in action. The north pole of one magnet will *attract* the south pole of another but will *repel* the north pole! Try suspending a bar magnet with marked poles, as in Fig. 9, by a string. Hold a second magnet as shown and observe the action. Notice that the

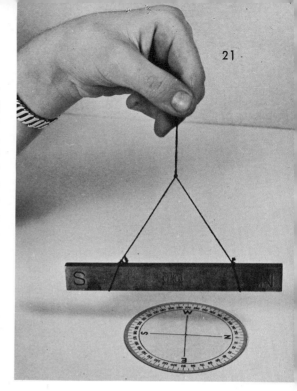

Fig. 8. Suspended magnets point north and south.

Fig. 9. Opposite poles attract. Free magnet moves toward fixed magnet.

Fig. 10. Like poles repel—hold themselves apart.

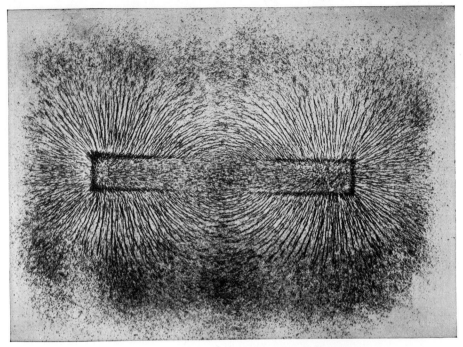

Fig. 11. Magnetic field around a magnet.

closer the magnets are together the stronger the action. Remember then that:

Like Poles REPEL
Unlike Poles ATTRACT

FIELD OF FORCE

Around every magnet there is a *field of force*. It is invisible, but we can prove it is there. If iron filings are sprinkled upon a piece of paper placed over a magnet, the filings will arrange themselves to show the direction of the lines-of-force of the "field" about the poles. See Fig. 11.

JOB 1. CHARTING THE LINES OF FORCE OF A MAGNET

Purpose: To show the lines of force of a bar magnet, Fig. 11.

Steps:

1. Obtain one sheet of 8½ x 11 drawing paper.

2. Print your name with ink in the lower right-hand corner.

3. Place this sheet in a flat pan of melted paraffin so that the sheet is entirely covered. Remove at once and let the extra paraffin drain off back into the pan. It will cool and appear dry. CAUTION: *Paraffin will burn rapidly. Do not let it get too hot in the pan. Turn the flame off while dipping.*

4. Place bar magnet on a flat surface with cooled sheet of wax paper over magnet.

5. Sprinkle iron filings over the paper and tap lightly with a pencil until you have a well-formed picture of the magnetic field.

6. Lift paper straight up away from the magnet and hold above an electric plate or another source of heat. Great care should be taken not to disturb the field.

7. Hold over heat until wax melts, or until the paper looks moist. Remove carefully and allow wax to cool.

The space about the magnet where magnetic forces exist is called the *magnetic field*. Notice that each filing turns in the same direction, an action caused by the magnetism of the two poles.

The lines of force flow out of the north pole of the magnet and back into the magnet at the south pole. The movement of the lines of force in the field is called *flux*. (Fig. 11) Flux, then, refers to the *flow of* magnetism as the word *current* refers to the *flow of water* in a stream. The stronger the flux, the stronger the magnet.

JOB 2. EXPERIMENTING WITH PERMANENT MAGNETS

Purpose: To learn more about magnets.

Steps:

1. Obtain a bar magnet, iron filings, a compass, a piece of writing paper, and miscellaneous pieces of material for testing, such as wood, iron, copper, cloth, etc.
2. Try the magnet on several kinds of materials. Notice which materials are attracted to the magnet.
3. Now hold a small needle compass near one end of the magnet. Then try the other end of the magnet; the needle is disturbed by the lines of force.
4. Place a piece of wood or cardboard over the magnet and sprinkle filings on the top surface to see if the lines of force go through such materials.

Magnetism will pass through almost any substance. For example, air, glass, and paper allow magnetism to pass through them and are said to be *magnetically transparent*.

There are substances which are highly magnetizable, such as iron, iron alloys, cobalt, and nickel. There are materials which cannot be magnetized, such as paper, wood, copper, and many others. These are called nonmagnetic materials.

MOLECULAR NATURE OF MAGNETISM

A molecule is the smallest particle of matter that can exist in a natural state. Molecules are made up of atoms. A molecule is so small that it cannot be seen through a powerful microscope. You will wish to learn more about both molecules and atoms later and also about ions, electrons, and protons. All substances are made of molecules. A small piece of steel, for example, consists of millions and millions of molecules of a type different from those in a piece of wood, or even in a different type of steel.

Molecules have electric charges. If a magnetized bar is heated red hot, it will lose its magnetism. Also, if the magnet is violently jarred or hammered, it will lose its magnetic strength. Such action is due to the disturbances of the molecules and their electric charges.

If a magnetized bar is broken, each part will be a complete magnet; two new poles will appear at the point of breaking, a new south pole on one piece and a new north pole on the other. We can continue to break each piece indefinitely, with the same result. See Fig. 12.

The molecules of a magnetized bar are little magnets lined up in rows, with their opposite poles in contact. All "N" poles face toward the north end of the bar, and all "S" poles face toward the south end of the bar. See Fig. 13. Disturbing this arrangement

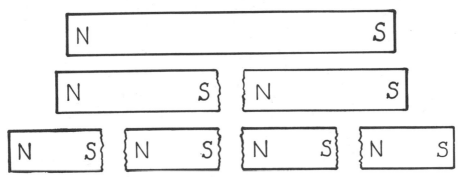

Fig. 12. Pieces of a broken magnet become smaller magnets without magnetizing.

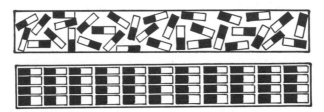

Fig. 13. The molecules in a magnet are small magnets.

by heat, by jarring, or by influence of an alternating current results in an unmagnetized bar, because the relation of the molecules or tiny magnets of which the bar is made is broken.

Until recently, it was believed that only steel or iron alloys could be easily magnetized. Today, such other materials as beeswax and plastic have been made into magnets. If material of this kind is placed in an intense electrostatic field while it is in liquid form and then cooled while still in the field, the material becomes magnetic. The electrostatic field arranges the molecules in the liquid material, as illustrated in the lower half of Figure 13, and these molecules are held in that position even after the liquid has changed to a solid.

THE EARTH AS A HUGE MAGNET

As previously stated, a bar magnet, held by a string so that the ends of the magnet are free to turn in any direction, will swing to a position pointing approximately north and south.

A compass is a magnet balanced on a needle point so that it will swing freely. Why does the compass needle point north? The earth is a large magnet, with magnetic poles near the geographical north and south poles. Your compass needle points toward these magnetic poles as it would be attracted to the poles of a bar magnet. Electricians, knowing that opposite poles attract, realize that the "north pole" of the earth, to which the magnetic north pole of a compass is attracted is, in reality, the magnetic *south* pole of the earth magnet. Do not be confused by this conflict of terms or wonder if there is an exception to the rule that opposite poles "attract." Simply re-

MAGNETISM

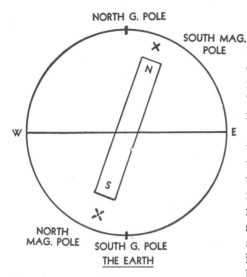

THE EARTH

The magnetic poles are many miles from the geographic poles.

member that the north end of the compass *points north*. It is so termed to make it easier to remember.

LOCATING GEOGRAPHIC POLES

The magnetic poles of the earth are not at the same location on the earth's surface as the true axis or *geographic* poles around which the earth turns. In fact, they are nearly 1400 miles at variance. A compass, then, does not point to the exact geographic poles but to the magnetic poles, and pilots and mariners who use the compass to tell directions must make corrections in compass readings in order to go where they plan to go. This last fact was first discovered by Columbus on his voyage to America.

Again, in some localities, there are large underground deposits of iron ore which attract the compass needle, thus preventing it from pointing true north. Voyagers must take this into consideration, also.

MAGNETIZING BY INDUCTION

The word "induction" is a form of the word "induce." Your friend induces you to go to a movie; he "influences" you. One magnet may so "induce" magnetism into an unmagnetized piece of steel.

Iron filings or nails that cling to the poles of a magnet also become little magnets. They are magnetized by *induction*. A nail may be suspended from a magnet; then a second nail may be suspended from the first, and so on. If we take the bar magnet away, the nails will fall apart, showing that they were magnets only as long as they were in contact with the bar magnet. See Fig. 14.

Any piece of soft iron may be magnetized only *temporarily* when it is held in contact with a permanent magnet or placed within a strong magnetic field. Yet, if the induced magnetism in the nail is tested with a compass the nail will be found to have *polarity*. There is opposite polarity on each nail.

Fig. 14. Magnetic induction.

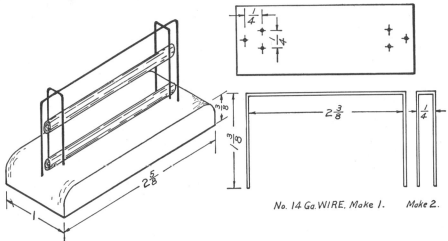

Fig. 15. The floating magnet.

You may wish to try this experiment in polarity to prove this fact to yourself. Test an unmagnetized nail with a compass before and after induction.

EXTRA JOB A: TO MAKE A FLOATING MAGNET (Fig. 15)

Steps:

1. Obtain wood base.
2. Chamfer or round top ⅛ inch. Sandpaper and shellac.
3. Mark and drill as shown.
4. Bend No. 14 wire as shown in the drawing, Fig. 15.
5. Obtain hard steel bars, 3/16" dia. x 2", magnetize, and place in holder, like poles together.

The bars may be magnetized by placing in a large spool of bell wire, the ends of which are connected to 30 V direct current or to a storage battery.

EXTRA JOB B: TO MAKE A PERMANENT MAGNET WITH A COIL OF INSULATED WIRE

Purpose: To show that steel may be magnetized from a direct electric current. See Fig. 6.

Experiment:

1. Obtain about 2 feet of bell wire and wind it about a pencil, making two or more layers and the coil about 2 inches long. Keep the two ends free and about 4 inches long.
2. Remove the insulation for about ½ inch from each lead end.
3. Place a needle in the coil after testing the needle with a compass to be sure it is not a magnet already.
4. Touch the ends of the lead ends to a dry cell for about ten seconds.
5. Test the needle for polarity with a compass.

Fig. 16. The cork-and-needle compass.

MAGNETISM

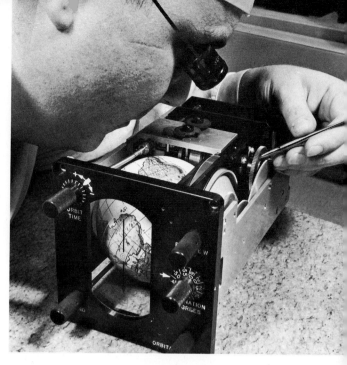

A technician checking an astronaut's space compass. Note the "sight" that shows his position in flight over the tiny globe compass.

COURTESY, HONEYWELL

EXTRA JOB *C*. To Make a Simple Compass

Steps:

1. Obtain a ¾ inch cork, magnetized needle, candle, and glass plate. See Fig. 16.
2. Slice off a disk ⅛ inch thick from a small cork. Lay a needle across the center of one surface and allow one drop of candle wax from a lighted candle to fall over the center of the needle. When cool, float the compass in a plate full of water and watch the action.

QUESTIONS

1. What is the difference between a loadstone and a permanent artificial magnet?
2. Will lines of force travel through glass?
3. Will soft iron (mild steel) make a better permanent magnet than high carbon steel?

For the Learn to Spell exercise, do the one on p. 16 of the Workbook.

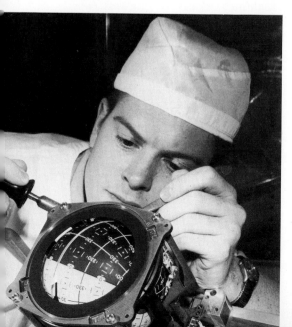

An astronaut's attitude indicator. The floating ball shows any error in flight of a space craft, such as *roll* (twisting), *pitch* (up or down), or *yaw* (veering off in an undesired direction). These technicians are highly trained.

COURTESY, HONEYWELL

Chapter III

SOURCES of Electricity and Electrical Energy

Two types of electricity—static electricity—current electricity—direct and alternating current—generators—electricity in cells—series and parallel wiring—voltage and amperage—symbols.

IT IS not easy to answer the question, "What is electricity?" Probably the simplest way to describe electricity is to say that it is a form of energy. Much has been learned about the effects it produces and about how it may be controlled. Generally speaking, there are two types of electricity: *Static* or *stationary electricity*, and *current* or *electricity in motion*.

STATIC ELECTRICITY

Static electricity was the first kind of electricity mentioned in history and little was known about it until the kite experiment made by Benjamin Franklin.

Static electricity is generated by friction. Friction of your shoes on a rug in dry, cold weather may generate a charge that will cause a spark to jump from your fingertip when you touch a radiator or another person. A leather belt driving over a pulley on a drill press has been known to generate sufficient static electricity to draw the unprotected hair of a woman operator into the pulley. It has blown up flour mills by igniting the dust, interfered with radio reception, stopped printing presses by pulling printed sheets into the wrong places, and knocked bridge toll collectors down when an automobile passenger handed them silver money. As for the demon, lightning, he has given no end of trouble.

Lightning strikes the nearest best conductor between the sky and the ground. Remember this in stormy weather, when you are in an open field, or on a mountain above the tree line, or standing in a boat.

Ben Franklin proved that lightning was static electricity generated by moving molecules of moisture in the clouds and discharging to other clouds or to an earthly object. Lightning rods have been erected on buildings and chimneys to take the static charges directly into the ground rather

Fig. 17. Some bridges have static discharge wires in the roadway, where toll is collected.

than have them pass through the structures.

Static, long a nuisance, is now being used. There is a machine which duplicates printed copy by applying an electrostatic charge to those areas of paper which are the images of lines, drawings, or type characters. The charge attracts dry, powdered printer's ink to the charged areas and holds the ink firmly until it is heated, melted, then cooled very rapidly. Electrostatics are also used in applying sand and other abrasives in the manufacture of sandpaper. Paint is sometimes applied to metal by static electricity.

CURRENT ELECTRICITY

You have learned that electricity, to be useful, must be controllable and not wild like a beast. Current electricity can be tamed and be made to do work for man.

The first person to produce an electric current was Alessandro Volta (1799). He built the first "dry" cell, which was made like a sandwich from a disk of copper, a layer of cloth moistened with salt water, then a disk of zinc, another layer of cloth, another disk of copper, and so on for several layers of the three materials. It was called the "Voltic pile." Wire connections were made to the first disk of copper and to the last disk of zinc. This connection produced current electricity. The word "volt" comes from Volta, name of the inventor. Our present dry cells and batteries of cells work on the same principles as the Voltic pile.

Two words whose meaning you must know are voltage and amperage, or volts and amperes.

Voltage means "force" or "pressure" of an electric current. Before electricity will flow, there must be force or voltage just as there must be force to move water. Gravity forces water down hill to make a current, and flowing water can be harnessed to do work.

Amperage is the volume of current —the measure of electrons flowing through a given length of a conductor in a given time.

A dry cell, when new, contains

about 25 amperes supply of electricity at a force or voltage of 1½ volts.

The word *battery* means a group of cells; several dry cells when connected form a battery. The cells in the car storage battery are 2 volts. Since the cells are connected in series, the 6-volt battery has three cells, and the 12-volt battery has six.

Amperage is the volume of current, the number of electrons flowing through a given length of conductor in a given length of time.

DIRECT CURRENT

The pressure from cells or batteries produces a current. The electricity flows out of one terminal, or contact, and back into the other. One terminal is called the *positive pole*. The other terminal is called the *negative pole*. The symbols for the poles are the + and − signs. We know that an electric current flows and, in the past, we thought of electricity as flowing out of the center pole of a dry cell and back into the outside pole. In recent years, it has been proved that the electrons flow in the opposite direction —out of the outside pole of a dry cell and back into the center pole. This means that the positive pole brings current from the negative pole. In this revised book, all symbols indicate the true direction of flow.

All unused cells produce a flow of current always flowing in the same direction from the − to the + poles. Such current flowing in one direction is called *direct current* (D.C.).

Mention has been made that both dry cell and storage batteries are sources of direct current.

Battery current is not used where heavy current is required for long periods of time. The conventional submarine is one exception, since it is one of the largest users of batteries. A few atomic-powered submarines rely on batteries for special services.

GENERATORS

The *direct current generator,* a machine for producing a voltage of electricity, is the cheapest and best means yet found to produce electricity on a large scale. Here *mechanical energy* is changed into *electrical energy.* All generators work on the principle of cutting the magnetic lines of force with coils of wire to collect the electrical charges.

A direct current generator produces direct current. A second general type of generator is the *alternating current generator* which produces an alternating current. To alternate means to move back and forth. A lion in a cage alternates from one end of the cage to the other, possibly all day. When he goes from one end of the cage to the other and then returns he completes one *cycle.* One cycle per second is called a hertz, a round trip a second. Imagine a light bulb attached to two wires a block long which are connected to an alternating current generator. The current first leaves the generator, going out on one wire, "A", to the bulb and returning to the generator on the other wire "B", then it goes out on "B" and returns on wire "A". Again it reverses itself, etc. Each round trip is a cycle, and a sixty-cycle current makes sixty cycles or sixty round trips *per second!* The polarity of the current in these wires changes 120 times per second.

SOURCES OF ELECTRICITY

Later, in Chapter XI, you will learn more about generators. For the present, let us go back to the electric cells and batteries to see how they produce electricity.

ELECTRICITY IN CELLS

It has been said that a molecule is composed of atoms. Each atom has an electric charge called *electrons*, which move around in the atom. A simple cell may be made by putting a strip of copper and a strip of zinc in a glass of water, adding a 10% solution of sulphuric acid. The chemical action works on the zinc atoms faster than it does on the copper atoms, releasing electrons. Electrons in motion produce a current; thus a direct current may be obtained by attaching terminals to the two strips.

If a wire is connected from each strip to a galvanometer (an instru-

Fig. 18. A simple cell connected to a galvanometer.

ment for measuring a small movement of current) a small current flow will be shown to be present. See Fig. 18.

A dry cell is composed of four main parts. See Fig. 19, and name them.

Fig. 19. Parts of a dry cell.

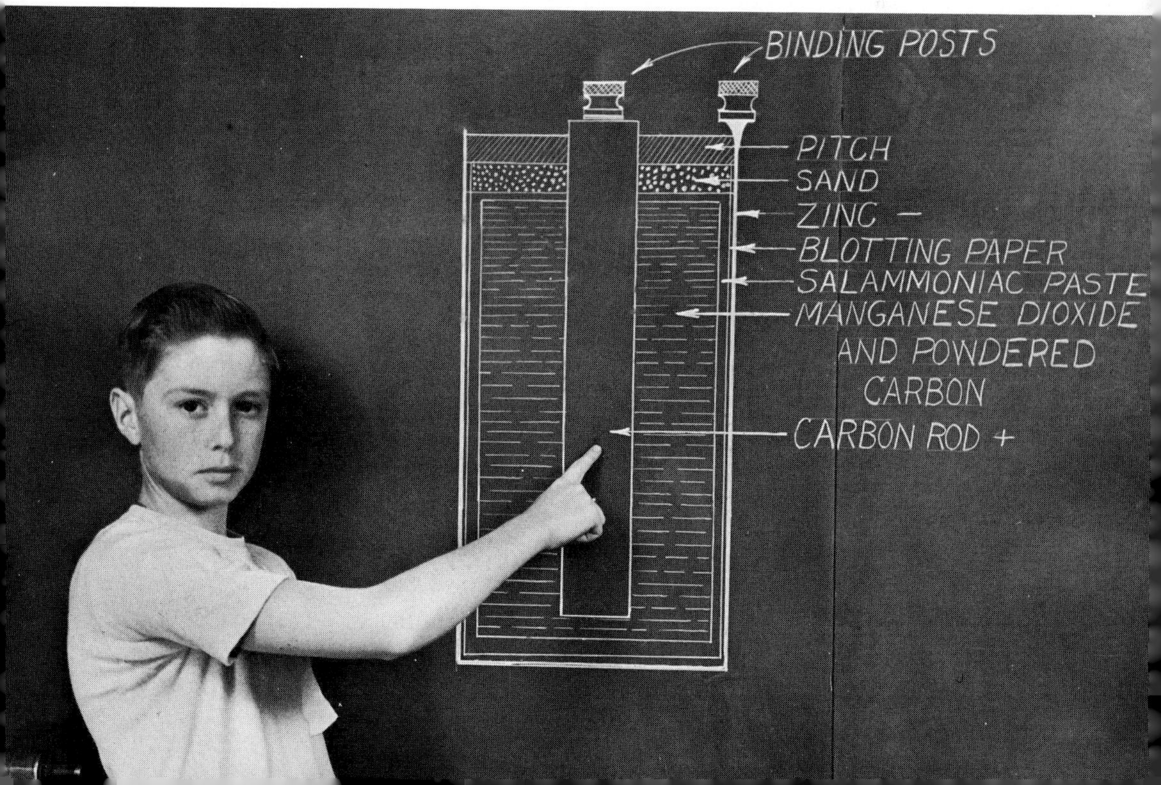

1. The zinc container.
2. The blotting-paper lining.
3. The chemical compound (electrolyte).
4. The carbon rod center.

The chemical compound is made of powdered carbon, manganese dioxide, sal ammoniac, and zinc chloride, which terms need not be memorized in this course.

Notice the two terminals or contacts. One is attached to the carbon and one to the zinc can. The carbon is the + or positive pole and the zinc is the − or negative pole.

SERIES AND PARALLEL WIRING OF CELLS

Since a battery is formed by connecting several cells, it is important to know how these cells should be connected to obtain the desired voltage for a circuit. Remember that the ordinary dry cell has a force of 1½ volts and its full capacity is about 25 amperes.

There are two common ways to connect cells, either in parallel or in series. (See Ch. III in Workbook.)

Suppose that your only source of light by which to read at night were a dry cell attached to a 1½ volt flashlight bulb hung over your bed. After several hours, the dry cell would be "dead"; that is, all the amperes would be used up. Suppose, then, you used four dry cells instead of one, so that you would need to replace the "dead" ones only a fourth as often. You would connect these in parallel, as shown in Figure 20. All the negative poles are connected to one wire that leads to the bulb and all the positive poles to the second wire.

Parallel wiring of several cells, then, gives only approximately the same force or voltage as one cell (see Fig. 20); but it produces approximately the combined amperage of all the cells. See Fig. 23.

Series wiring is illustrated in Fig. 21. Note that one lead comes from the negative pole of the last cell in the line. All other poles are connected likewise, the negative of the first to the positive of the second, and so on.

Now the amperage or amount of current here cannot be any greater than that in the parallel circuit, because the same number of cells are used; but the voltage or force which drives the current out is four times as great, or six volts, since there are four cells and each cell is 1½ volts. Fig. 23 will help you remember in which

Fig. 20. Cells connected in parallel.

Fig. 21. Cells connected in series.

COURTESY DUNWOODY INSTITUTE

Communications laboratory for training technicians.

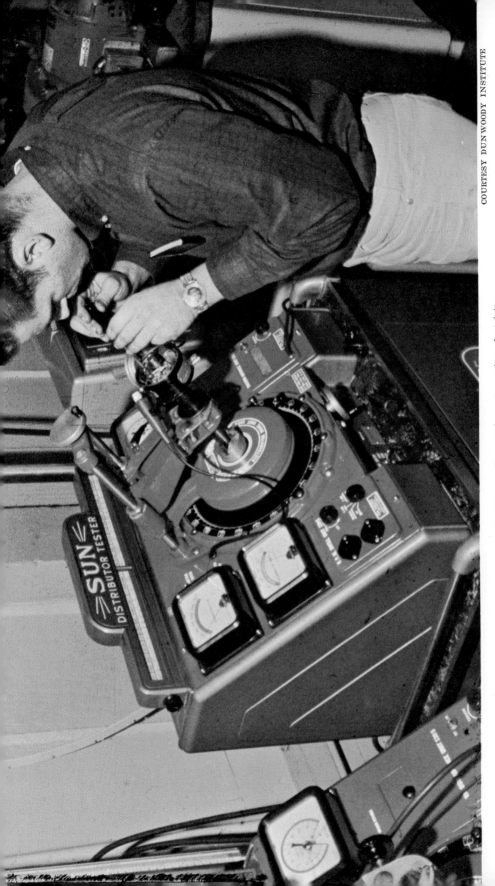

Testing equipment for automotive electricity.

COURTESY DUNWOODY INSTITUTE

SOURCES OF ELECTRICITY

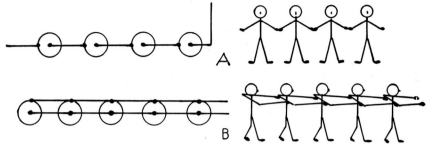

Fig. 22. The *force* is greater when cells are connected in series than in parallel.

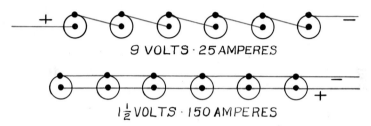

Fig. 23. Six cells in series and in parallel.

type of connection you get the more force, in parallel or in series connections.

Now if you were to connect your bed light bulb to the cells when connected in series, you would be forcing electricity with six volts through a 1½ volt bulb. The bulb would "burn out" or "go dead." A six-volt bulb would be necessary to hold six volts of pressure.

In the game "crack the whip," all players join hands in series (as in "A," Fig. 22) to get a great deal of *force* on the player at the end of the line: but in "parallel," as in "B," the force is expended upon each player.

For years, the dry cell and a few sizes of flashlight batteries were the only type of cells available. However, through research, a large variety of such cells have been developed. Here are a few; can you name others?

• The manganese alkaline cell for firing flash bulbs when using a camera.
• The nickel cadmium rechargeable cell for powering cordless electric razors, etc.
• The tiny nickel silver cell with a lifetime warranty to light a keyhole.
• Transistor cells of various sizes and voltages.

JOB 3. EXPERIMENTING WITH DRY CELLS

Purpose: To learn to connect cells.
Experiment:

1. Obtain four used dry cells, a voltmeter, 0-25 volts, an ammeter, and about 3 feet of No. 18 annunciator or bell wire.
2. Arrange the cells for wiring in series.
3. Remove insulation where needed and connect; leaving two lead wires to run to the voltmeter.
4. Find by arithmetic how many volts should be shown on the meter.
5. Connect to meter and check.
6. Do the same for testing amperage.

MEASURING VOLTAGE AND AMPERAGE

It has been said before that voltage is the pressure that causes a flow of electrical energy or the *current flow*. Without voltage or force, electricity would not move. Voltage is measured with a *voltmeter* somewhat as a steam gauge measures the steam pressure on a boiler. *Great care should be taken when using any electrical measuring instrument.* See that the instrument you use will measure *more* than the required amount and also that the polarity is correct. A voltmeter should always be connected in parallel or across the load. See Fig. 24.

André Marie Ampere was a French scientist who first gave us real information on electric current. Current flow, consequently, is named after Ampere—amperage or ampere. The ampere, then, is a measurement of current flow, or the amount of current flowing past a certain point in a given time, and is measured just as a water meter in your home measures all the water passing through a pipe into your home.

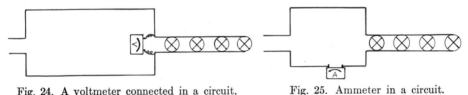

Fig. 24. A voltmeter connected in a circuit. Fig. 25. Ammeter in a circuit.

Fig. 26. Types of meters.

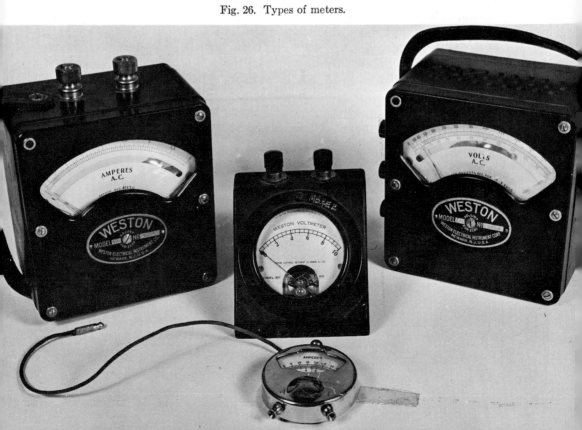

SOURCES OF ELECTRICITY

The ammeter is an instrument to measure the rate of current flow. The ammeter must be connected in series or into the line so that *all* the current at that point must pass through it. See Fig. 25. Fig. 26 shows a small combination meter (front) which measures both voltage and amperage of dry cells or other low-voltage sources.

SYMBOLS TO BE LEARNED

In the electrical field, men work from plans and drawings. On these plans appear certain symbols which tell them where to place a certain device. In making your plans or drawing, you will use symbols as they are needed. As you study each unit in this text, you will find new symbols to learn.

For this unit, learn the following symbols and abbreviations (see illustration):

D.C. = Direct Current
A.C. = Alternating Current

─┤+├─┤−├─ = CELL

─┤+├┤├┤├─ = BATTERY

⌐V⌐ = VOLTMETER

⌐A⌐ = AMMETER

EXTRA JOB *A:* Cut a dry cell down the center with a hack saw. Clean parts, mount on a panel, and label.

EXTRA JOB *B:* Read about static electricity and make some pith balls from cornstalk centers or from puffed wheat kernels. Mount on a stand with thread. Experiment with a comb and the pith balls to show varied charges. Fig. 27.

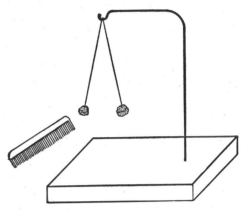

Fig. 27. A static experiment.

EXTRA JOB *C:* Make a simple electric cell from a lemon, as follows, and demonstrate the experiment before the class. Obtain a piece of copper and a piece of zinc, each about ½″ wide by 2″ long, and wind a piece of No. 28 insulated wire around a 1¼″ box nail. Solder one end to one corner of the copper and the other end to one end of the zinc. With a knife, cut two slits into the lemon, each about ½″ long, as shown below, and insert the zinc into one slit and the copper into the other.

Electric energy from a lemon.

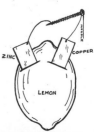

QUESTIONS

1. What type of electricity causes a small piece of paper to be attracted to a hard rubber comb after the comb is rubbed on a piece of wool or fur in cold, dry weather?
2. What is the voltage of the line that comes into a home for lighting and power?
3. If a generator produces electricity when being driven by a motor, why not run the motor from the power generated and have "perpetual motion"?
4. A boy connected a 110 volt source to a 15 volt voltmeter. What was the result?
5. If lightning is the static discharge of a cloud, what is thunder?
6. Does one see a lightning flash before or after hearing the thunder clap? Why?
7. What is the voltage of three dry cells in series? See illustration.

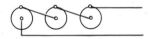

8. What is the voltage of three dry cells connected in parallel? See illustration.

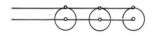

9. What is the amperage of five dry cells connected in series? See illustration.

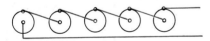

10. What is the amperage of five cells in parallel? See illustration.

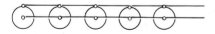

METER READING EXERCISE

EXTRA JOB *D:*

1. Obtain four used dry cells, a voltmeter, 0-10 volts, and an ammeter, 0 to 100 amps.
2. Number the cells I, II, III, and IV.
3. Down the left margin of a piece of paper, number from 1 through 20, allowing space after each number for one answer each.
4. Using the equipment above, fill in your answers for the following:

   a. What is the voltage of Cell I?
   b. What is the voltage of Cell II?
   c. What is the voltage of Cell III?
   d. What is the voltage of Cell IV?
   e. What is the amperage of Cell I?
   f. Of Cell II?
   g. Of Cell III?
   h. Of Cell IV?
   i. Connect two cells in series: What is the combined voltage? What is the combined amperage?
   j. Connect three in series: What is the combined voltage? What is the combined amperage?
   k. Connect four in series: What is the voltage? What is the amperage?
   l. Connect two in parallel: What is the combined voltage? What is the combined amperage?
   m. Connect three in parallel: What is the voltage now? What is the amperage now?
   n. Connect four in parallel: What is the combined voltage? What is the combined amperage?

LEARN TO SPELL

electron	negative
current	positive
chemical	series
parallel	ohm
amperage	voltage
friction	static

Chapter IV

ELECTROMAGNETICS or Magnetism Induced by Electrical Flow

How the electromagnet is formed—the solenoid or coil with movable core—induction coil—electrical "shock"—induction coil, voltage, spark coil, and transformer.

THE EFFECTS of the flow of electricity are interesting. Suppose that you are holding an insulated wire in your hand which is the return wire of a direct current circuit operating a small motor or a light or a bell. Going around and around that wire on the *outside* of the insulation, there are magnetic lines of force like those you found about a permanent magnet. You cannot see the lines of force or feel them or hear them, but they are there; and they will be there until the current is shut off.

THE ELECTROMAGNET

Around any electrical conductor through which a current is flowing, there are magnetic lines of force. Fig. 28 shows the lines as the iron filings arrange themselves on a piece of cardboard through which a wire runs to and from a cell. Notice the three compasses which show the polarity in three different places.

Also, observe that the filings show greater attraction close to the wire.

It was these "magic" lines of force in the coil which you made in Chapter II that magnetized the needle. A coil of insulated wire wrapped around a bar of mild steel (called a core) becomes an "electromagnet" when the ends of the coil are attached to a direct current. Electromagnets are more commonly used than permanent magnets. They may be found in tele-

Fig. 28. Magnetic lines around a wire flow into the south end of the compass needle and out of the north end.

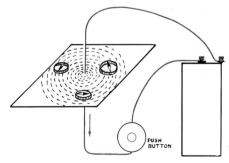

A

COURTESY, D. ONAN & SONS, MINNEAPOLIS

phones, telegraphs, radios, doorbells, relays, motors, etc. Fig. 29 shows a huge electromagnet sufficiently strong to pick up steel from the highways. Powerful magnets are used to lift scrap iron and other forms of steel.

Keep in mind that the steel core on the inside of the electromagnet must be mild steel, commonly called soft iron. A hard steel holds mag-

Fig. 29. Large electromagnets used for different purposes. (A) Road magnets. (B) Roller skating on a lifting magnet. (C) A lifting magnet at work.

B

C

COURTESY, DINGS MAGNETIC SEP. CO.

COURTESY, CUTLER-HAMMER, INC.

ELECTROMAGNETISM

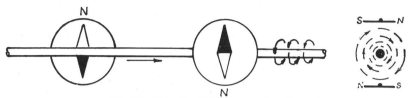

Fig. 30. Polarity of lines around a wire.

netism after the current is turned off. Mild steel loses its magnetism when the current is turned off.

The next job for you to do is an experiment with a compass and an electric current. See Fig. 30.

JOB 4. TO SHOW LINES OF FORCE ABOUT A WIRE

Purpose: To prove that there are magnetic lines of force around a conductor and that these lines have polarity.

Steps:

1. Obtain 3 feet of bell wire; a dry cell; a piece of cardboard about 12" x 12"; iron filings; a compass; a screw driver.
2. Wire a circuit as in Fig. 31.
3. Lay the wire down on the bench.
4. Hold a compass *under* the wire, close the circuit, and note which way the needle points.
5. Now hold the compass *over* the wire, close the circuit, and watch the action of the needle.
6. The needle will be at right angles to the wire, but it will point in the opposite direction from which it pointed when it was under the wire. *The north end of the compass needle points in the direction of flow of the magnetic lines of force.* If you reverse the direction of the current by connecting the wires on the other poles of the cell, the compass needle will reverse also.
7. Now form the wire into a loop and check the lines of force at various places on the loop. See Fig. 31.

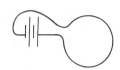

Fig. 31. A loop of wire in a circuit.

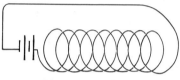

Fig. 32. The helix.

8. Wind the wire around the handle of a screw driver or around a 1" dowel to form it into a series of loose loops about ¼ inch apart. See Fig. 32.
9. Check the polarity of the loop ends. One will be a north pole, the other a south pole. If you grasp the coil with the right hand with your fingers going in the same direction as the current is flowing in the loops, your thumb will point to the north pole of the coil. Check this statement.
10. Slide the blade of the screw driver into the coil and note that the coil attracts the compass with greater strength than a coil which has no core.
11. With a compass, check several screw drivers to find one which is not a permanent magnet. Hold it inside a coil of wire such as the one above. Turn on the current for about twenty seconds. Take out the screw driver and note that it has become a magnet.

JOB 5. To Make an Electromagnet

Purpose: To learn how to make an electromagnet to be used later in the telegraph project. See Fig. 33.

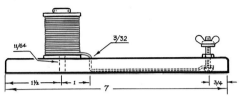

Fig. 33. A mounted electromagnet.

Steps:

1. Obtain a spool of No. 24, enamelled, cotton-covered wire; one 20d nail cut off to be 2 inches long; two 1-inch fiber washers; one piece of ½-inch pine 3″ x 7″; two stove bolts, nuts, and wing nuts.

2. Drill the five holes in the base. See Fig. 33. Then groove out on the bottom for the two wires.

3. Drill a 3/16-inch hole in the center of each fiber washer; drill two 1/16-inch holes in the lower washer as in Fig. 34; then force the washers onto the 20d nail, spacing them 1½ inches apart.

4. Wind the wire about the nail in very tight turns and layers. See Fig. 34. After finishing the last layer, cut the wire to allow six inches of lead to be pushed out of the outer hole.

Fig. 34. Winding the magnet.

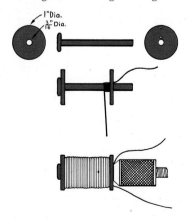

5. With both hands, press the two fiber washers toward each other and ask someone to loosen the chuck. Then immediately place the nail in the base hole and tap the coil down in place.

6. Connect the wires to two binding posts.

7. Connect a dry cell and test for magnetism.

8. Keep this magnet assembled so that you can use it later in making a telegraph sounder.

9. Return the other materials.

THE SOLENOID

A solenoid is an electromagnet, the wire coil of which is wound on a thin walled brass tube. A core of soft iron fits loosely into the tube and can be moved in and out of the coil. This core is called the armature. The coil, when connected to a direct current source, becomes a magnet and sucks the armature into its center. The pull of the coil on the armature is a feature which makes the solenoid a useful machine for turning valves and switches on and off. The home automatic washer uses several solenoids.

Fig. 35a shows the principle of the simple door chimes. Fig. 35b shows parts. The core "D", Fig. 35a, passes through the hollow center of the coil "C". This core projects from the lower end of the coil and rests on shelf "F". When button "B" is pressed, the current from one terminal of the cell goes through the switch, then through the coils to return to the cell. The magnetism in the coil quickly draws the core "D" off the shelf. Then the core jumps out the other end by its own force to strike against the bell-metal strip "E", which gives the alarm. When "B" is released, the core falls back on the shelf.

THE INDUCTION COIL

The induction coil is really two coils of wire wound together but connected separately. The simplest induction coil is illustrated in Fig. 36. Here you see a coil of Number 18 insulated wire wound about a core, with its two ends connected in series with a bell and a dry-cell circuit. On the outside of the first coil is wound a second coil of fine wire (No. 32) with its ends connected to two old carbon centers from a worn-out dry cell. The inside coil is called the *primary coil*, and the outside coil, of small wire, is called the *secondary coil*. When the switch is closed, the current flows through the primary coil but is interrupted each time the bell arm leaves the contact for the magnet in the bell. At each "interruption" a strong flow

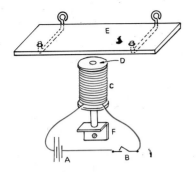

Fig. 35a. The mechanism of the small chimes.

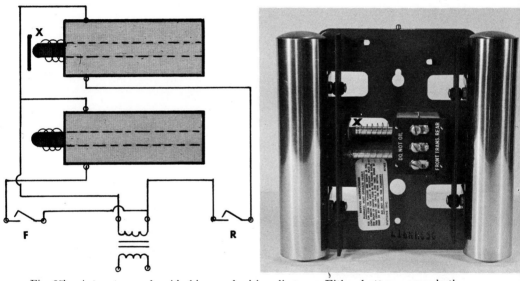

Fig. 35b. A two-tone solenoid chime and wiring diagram. Either button powers both solenoids, but only the lower striker hits both left and right chime tubes. See the stop at X for the rear door single tone.

Fig. 36. The induction coil in a circuit.

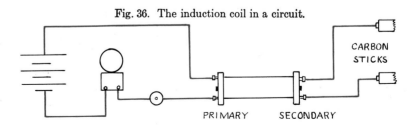

of current is induced into the secondary coil. In fact, the voltage is so great that you get a shock by touching the two carbons. A strong primary current may even induce a voltage in the secondary coil sufficiently strong to make a spark jump between the carbons if the carbons are placed $\frac{1}{8}$ inch apart.

JOB 6. SIMPLE INDUCTION COIL

Purpose: To make and use an induction coil.

Steps:

1. Obtain a buzzer or door bell; dry cells; two pieces of carbon from old dry cells; stove-pipe wire; No. 18 bell wire; No. 32 magnet wire; and two pieces of lamp cord each about 2 feet long.
2. Cut eighteen pieces of soft iron wire 6 inches long. Any soft iron wire that will make a core about $\frac{3}{8}$ inch to $\frac{1}{2}$ inch in diameter may be used. Form these into a bundle or core. If no wire is available, use a $\frac{3}{8}$-inch carriage bolt, 6 inches long.
3. Cut two pieces of $\frac{1}{2}$-inch pine in 2-inch squares.
4. Draw diagonals on face and bore a hole through each block. This hole should be just large enough to admit the core and hold it securely.
5. Bore two $\frac{1}{16}$-inch holes in block "A", in positions shown on the drawing, Fig. 37, and two in block "B".
6. Insert the wire core in the two blocks, allowing one end of core to project $\frac{1}{4}$ inch out of one block, to be clamped in an auger brace or drill chuck for winding in the wire coils. See Fig. 37.
7. Wind four layers of No. 18 wire, allowing 1 inch of each end of the wire to protrude from the hole in block "A". Secure the ends on the outside of the block with thread. Count the turns of wire.
8. Wind about a $\frac{1}{2}$ inch of the fine wire about the primary. If you have a coil winder, count the turns. The ends of this wire protrude from the holes in Block "B".
9. Cover the coil with cloth, and shellac.
10. Fasten two Fahnestock connectors, or roundhead wood screws $\frac{1}{4}$ inch long, to both blocks and solder the coil ends to them.
11. Connect as shown in Fig. 36.
12. In Fig. 37, notice the position of lead wire holes in block "B" for primary winding and block "A" for secondary winding. Use $\frac{1}{16}$ inch drill.

One end of core is $\frac{1}{2}$ inch out of one block, Fig. 37, so that this tip of the core may be clamped in a drill-press chuck or a speed-drill chuck. The small wire may be wound faster in this manner than if done by hand and also wound without kinking.

The bell was necessary in the above job to make and break the circuit. Many induction coils have their own "make and break" mechanisms which work exactly like the mechanism in a bell.

One other condition will cause a current impulse in the secondary circuit of an induction coil. If a loose core in an induction coil is pushed out or shoved in quickly, an impulse will be generated in each instance, provided there is a direct current in the primary.

Another method of inducing a con-

Fig. 37. One end of the core projects out for easy winding.

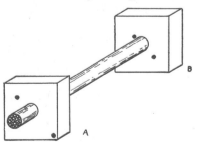

tinuous flow of electricity in the secondary coil is to run an alternating current through the primary. Since an alternating current first flows one way, then stops in order to flow the other way, each reversal of direction induces the secondary coil to pick up the magnetic flux or an impulse. If the primary circuit is sixty cycles, or, in other words, 120 changes of direction in flow per second, there will be 120 pulsations of secondary current per second.

SHOCK FROM AN ELECTRIC CURRENT

You may have noticed that when the induction coil was connected to a dry cell it gave off a "shock" or a high-voltage impulse each time the circuit was completed and that it was not a *continuous series of shocks* even though the current was still on. It was necessary for the circuit to be continually completed and broken before fresh shocks were felt! Electric impulse will flow, then, from the secondary coil only when and at the instant the current is turned off or on.

Can one get a shock from a dry cell? Not unless a change is made in the force of the current flowing from the cell. The dry cell has amperage and voltage, and it is the voltage that can be so increased that the current attempts to flow through your arms to produce a shock. An illustration of an effect similar to changing a dry cell of electricity to produce a shock is as follows: Suppose the empty, zinc, dry-cell container were filled with loose sand and that a small hole was punched in the bottom to allow the sand to flow out gently when held in the air. This stream of sand would hardly be felt by your hand if it were held in the stream below the container because of the lack of "force." But suppose something is done to change that force of flow, using the *same amount* of sand. If a stream of fast-flowing air from a filling-station air hose forces the sand against your hand until the sand stings the surface, you would be physically "shocked."

Induction Coil and Voltage

The number of turns of wire on an induction coil determines the change in voltage. For example, an induction coil with a good core has a hundred turns of wire on the primary circuit and is connected with a buzzer and a battery. The secondary coil has a thousand turns, or ten times as many turns as on the primary. A six-volt battery forces the electricity through the primary circuit. The secondary coil, having ten times as many turns of wire as the primary coil, has approximately ten times as many volts, or sixty volts. Now sixty volts or force is sufficiently strong to be felt by touching. In other words, anyone who touches both terminals of the secondary coil will receive a shock.

People differ in their "ability" to be shocked by a current of electricity. The nerves and the heart are affected by shock. A shock which might be fatal to some people would hardly be felt by others. Also, the electricity, in attempting to flow through the body, needs a good connection. Since water will conduct electricity, a person may get a severe shock if he is standing on a damp floor, or in a bathtub, or if the hands are wet when in contact with an electric appliance.

The Spark Coil

The strength of an induction coil depends upon the size of wires used, the number of turns of wire in each coil, the current used, and the workmanship of the person making the coil. A coil stronger than the one you made will produce an arc across the secondary terminal if the terminals are placed close together. Such coils are used in the automobile to send sparks into the cylinders to explode the gasoline and air mixture which drives the piston down to run the crankshaft.

Spark coils are commonly used in the automobile for the purpose of igniting the compressed gas in the cylinder. A used coil from an automobile junk yard can be connected so that it will throw a strong arc. You may wish to design and build a *shocker* similar to the one shown in Fig. 36.

The Simple Transformer

The simple transformer is a type of induction coil. The transformer "transforms" the current within a circuit, increasing or decreasing the voltage. When a transformer increases the force, the process is spoken of as being "stepped up"; "stepped down" is the term used for decreasing the force. It is interesting to study the construction of the core of a transformer. This core is usually made in the form of either a doughnut-shaped ring or a square block

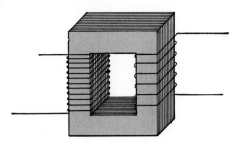

Fig. 38. Primary and secondary coils on a transformer core.

with a square hole through the surface. Fig. 38 is a drawing of a simple transformer. The primary coil is wound on one side, and the secondary is wound on the other. (A current flowing through the primary makes a magnet out of almost the entire square block.)

The secondary coil picks up the lines of force each time the alternating current is made or broken by its reversals of direction. The number of turns of wire in each coil determines the amount of transformed voltage.

Suppose a transformer is wound with a proportion of ten to one, in the number of loops. Suppose there are one hundred turns on one side and one thousand turns on the other side. If fifty volts of alternating current were connected to the coil with the hundred turns, the voltage in the other would be ten times that amount or five-hundred volts. Of course, there would be some loss of current, but the actual voltage would be almost five-hundred volts. Now, if the fifty volts were connected to the opposite coil, the voltage would be stepped down one tenth, or to about five volts. See Fig. 38. An alternating current is required for transformers.

ELECTROMAGNETISM

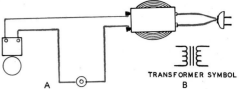

Fig. 40. The transformer used in a bell circuit.

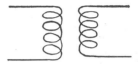

Symbol for a transformer

EXTRA JOB *A:* Find a good plan for a simple, easily made transformer. It takes many hours to build a good one. The 110-volt coil of the transformer probably should be wound and connected after you have studied the chapter explaining 110-volt wiring.

EXTRA JOB *B:* Design and build a model-railroad semiphore signal about 10 inches high. Make a solenoid, mount vertically, and suspend the core from the signal arm to drop part way into the hollow center of the coil.

EXTRA JOB *C:* Design a toy solenoid motor with a flywheel, crankshaft, connecting rod, and "piston" (piston here to be the movable core of the coil).

Fig. 39. Transformer mounted on panel for circuit wiring.

QUESTIONS

1. Why may a compass reading be unreliable when it is held in the front seat of an auto?

2. What would be the action of a compass held over an insulated, 110 volt, alternating-current wire, with the current flowing?

Transformer cores are usually made of many sheets of black, soft sheet steel put together in layers called *laminations*. Examine a small, bell-ringing transformer and look for the laminated core. See Fig. 39.

46 INDUSTRIAL-ARTS ELECTRICITY

3. Under what conditions will the current flow in the secondary circuit of an induction coil?

4. What do you think causes the burns on a person who has had a very severe shock from a strong electric current?

5. John buys a bell-ringing transformer to reduce 110 volts to 11 volts, so that he may replace the batteries in his doorbell circuit. He connects the wires improperly, so that the 110 volts enter the terminals which were supposed to be used for the bell circuit. How many volts would there be at the other two terminals?

6. Read the following: 1 transformer 110 V, A.C., 60 cy., 5-15 secondary, 15 watts, 50¢ ea.

7. A 10,000 volt A.C. power line runs near a home not wired for electricity. If the house were to be wired for 110 volts what instrument would need to be purchased before the house circuit could receive current from the power line?

8. What First Aid should be given one found unconscious from an electric shock?

9. Is every individual affected the same by a shock from an electric current?

10. What is an "electric fence?"

(See Work Unit No. 6 for Chap. IV in the Workbook.)

LEARN TO SPELL

solenoid	secondary	primary	coil
core	cycle	zinc	proton
transformer	armature	terminal	laminations

An electronics technician servicing a data processing machine which has a magnetically operated keyboard.

COURTESY, CONTROL DATA

Chapter V

THE FLOW OF ELECTRICITY and Conducting Materials

Conductors and insulators—electron theory, kinds of wire, insulation, splicing—symbols for wiring—switches—electric bell and buzzer—terminals—using dry cell or transformer.

FROM previous chapters, we learned that electricity travels through various substances and that this traveling is called *flow*. Substances which conduct the flow of current electricity or permit electricity to flow through them are called *conductors*. Substances which do not permit the flow of electricity are called *insulators*.

CONDUCTORS AND INSULATORS

Conductors vary as to their conductivity. In other words, there are good, medium, and poor conductors. Electricity must be controlled if it is to work for men. There must be a conductor (or path) on which electricity is to travel, and also there must be insulators to keep it from "leaking off" the conductor. All materials used in electrical wiring will come under these two headings: either *conductors* or *insulators*.

You have learned that all matter is made up of atoms and that each atom has positive and negative charges of electricity. The movement within a wire of these negative charges, which are called electrons, is interesting. If the electrons are easily moved, then we have a good conductor, such as in silver and copper. But otherwise, if the electrons are bound tightly within a material, such as in glass or mica, it is almost impossible to move them from atom to atom. The latter is true of a good insulator (see table below).

COMMON CONDUCTORS

silver	mercury
copper	carbon
aluminum	seawater
zinc, brass	moist earth
iron	the body
nickel	acid solution

COMMON INSULATORS

dry air	oils
glass	porcelain
rubber	silk
cotton	paraffin
paper	shellac

Fig. 41. Testing for conductivity.

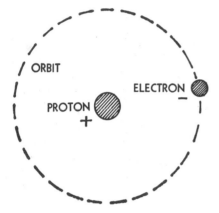

Fig. 42. The atom with its proton and electron.

To try several substances to prove their insulating qualities, or conductivity, hook a dry cell to a doorbell, but break one wire and hold the material to be tested to the two broken ends. See Fig. 41.

THE ELECTRON THEORY

It has been said before that matter is anything that has weight and occupies space; matter is made of small parts called molecules: and these molecules in turn are made of still smaller particles called atoms. It is the atom which is of interest here.

See Fig. 42. Each atom, it is assumed, has a nucleus called a *proton* having a positive (+) charge. Revolving about the nucleus are one or more electrons with negative charges. A normal atom has as many protons as electrons and is said to be equal in charge; but let an outside influence upset this balance, such as the chemical action of a dry cell or the voltage of a generator, and we force electrons to leave the atom and travel to the next atom, etc. Current electricity is the controlled flow of electrons.

WIRE AND WIRING

Kinds of Insulated Wire

The most common conductor used is copper in the form of wire. Silver is a better conductor than copper but is more expensive and less plentiful.

Copper wires vary in diameter and in the type of insulation, depending upon the use to be made of them. The diameter of the copper wire is measured with a wire gauge, as in Fig. 43. The smaller the gauge number the larger the wire. To select wire, consider the following:

1. What gauge should be used?
2. Should it be solid wire or stranded?
3. What type of insulation should it have?
4. How long should it be?

Kinds of Common Wire

The larger the diameter of a copper wire, the more amperage it will carry but the more costly it is. Electricians

Fig. 43. The wire gauge and spools of magnet wire and bell wire.

usually select the smallest wire that will safely carry the necessary amperage. If a wire too small for a job is selected, the wire gets hot, burns off the insulation, and may set fire to anything near by which might catch fire. The wire may also burn itself in two and thus break the circuit.

Insulation Materials

There are many kinds of insulating materials used on wire. The kind selected depends on the purposes for which the wire is to be used. Enamel, lacquer, or a layer of cotton thread—any of these are sufficient for magnet wire. Rubber or flexible plastic is popular for flexible extension cords. Solid wire, rubber covered, then covered with a sleeve of cotton fabric is approved for wires that run within the walls and floors. However, it must be enclosed in rigid, or flexible, metal tubing. Heating appliances must have asbestos-lined insulated wires. See Fig. 75, page 72. Wire for connecting door bells, or other low-voltage home circuits, is cotton wrapped and coated with paraffin.

Have you ever seen a newspaper story which starts something like this: "Fire Burns Home—Family Escapes —Cause Laid to Defective Wiring"?

Hold the wire at a right angle from the flat side of the wire gauge.

49

Such news items are still very familiar to all even though we have strict laws governing the wiring of homes and of other buildings where people live and work. The National Underwriters have formed a set of rules and standards for the manufacture of all electrical materials requiring dangerous installations. Most communities have additional laws which require that only electricians who are licensed may wire homes and buildings. Then, too, electrical inspectors employed by the communities check installations to reduce the number of hazards to a minimum, and yet with all this effort of society to see that those who use electricity do so safely, thousands of people are hurt or killed and property in the millions of dollars is lost annually by electricity "out of control."

You may find the name "National Underwriters" on a clip attached to approved appliance cords, showing that the material so labeled meets the correct standards. See if your community has local rulings and, if they are available, try to get a copy of them to read.

Splicing

Wiring consists of running conductors, usually wires, from the source of current to control devices such as switches, and to the equipment to be installed. When the wire at hand is too short to reach a required distance, it is necessary occasionally to join two pieces together. See Fig. 44 at "Y". The joint where the two wires are put together is called a *splice*. The most common type of splice is the Western Union splice illustrated in the Job instructions which you are to do later. See Fig. 45.

Fig. 44 shows the use of the tap joints at "X". You will learn in Job 8 to make one of these and to do the soldering. Most building electricians use patented connectors like those illustrated on page 95, rather than the soldered splice. In electronics, taps are commonly soldered to small half-inch long metal lugs, one end of which is either fastened to a common ground or to an insulated connecting strip.

If the splice or tap joints are made properly, the current will pass through them as though it were a solid piece of wire. But if the wires are joined loosely, the current may not pass through or may try to jump an open space and cause "arcing."

Arcing is a term which describes electricity as it *jumps through the air* from one conductor to another. Arcing is dangerous, because of the heat generated. Of course, arcing takes place only in currents strong enough to force the electricity to jump across an air gap. Joining wires which carry a strong current that is apt to arc requires that the workmanship meet certain standards. For your own safety and that of others, learn to connect wires properly.

Fig. 44. The tap and splice joints in a circuit.

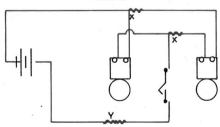

Splicing in Low-Voltage Circuits

In splicing, the joints must be as strong as or stronger than if there were no joints in the wire. To join wire, of course, the insulation must be removed in making the joint and the wire cleaned so that a good contact will be assured. When almost any conductor is left open to the air for a period of time, it tarnishes or oxydizes. This oxide of the material is a poorer conductor than the material itself and needs to be scraped off before connections are made to the material. After the joint is finished, the insulation must be replaced.

A good joint is one that is strong, will permit electricity to flow through it and do so without arcing, and one that is again protected with insulation. Wires that carry a heavy current need to have the joints soldered, which you will learn to do later. In this unit on low-voltage wiring you will not need to solder the splices.

JOB 7. WESTERN UNION SPLICE

Purpose: To make a Western Union splice in bell wire. See Fig. 45.

Steps:

1. Obtain two pieces of No. 18 bell wire about 8″ long each.
2. Remove the cotton insulation by unwrapping and cutting off or pushing back 2 inches from one end of each wire.
3. Polish the exposed wire by scraping lightly with the back of a knife blade.
4. Hold as shown in the diagram and twist with the fingers. Make two long wraps, then see that you have four or five tight turns at the ends about the wire. Have inspected by your instructor.
5. **Tape on** bell wiring jobs is not often

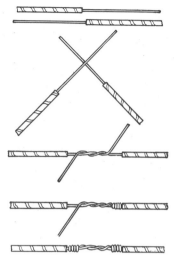

Fig. 45. The Western Union splice in bell wire.

used but, if required, wrap tightly with about 4 inches of electricians' tape.

6. Attach a tag with your name and hand in for credit.

JOB 8. TAP SPLICE

Purpose: To make a tap splice. See Fig. 46, page 52.

Steps:

1. Obtain two pieces of No. 18 bell wire about 8 inches long each, a knife, and a pair of pliers.
2. Remove about 2 inches of insulation from one end of the branch wire, "A".
3. Call one wire, the main wire, "B", and cut about 1 inch of insulation out of the center, being careful not to cut or nick the wire.
4. Clean the bare surfaces of both wires.
5. Wrap the branch wire around the main wire, starting as in "C".
6. Make two loose turns and four close turns as in "D".
7. Cut off the surplus wire and press the ends down smooth with the pliers.
8. Have inspected by your instructor before taping and hand in for credit.

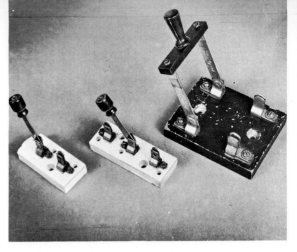

Fig. 47. Knife switches — the single-pole, single-throw; the single-pole, double-throw; and double-pole, single-throw types.

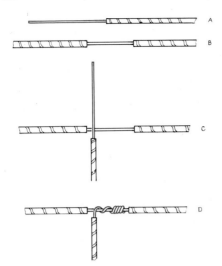

Fig. 46. The tap joint in bell wire.

SYMBOLS FOR YOU TO USE AND LEARN TO DRAW

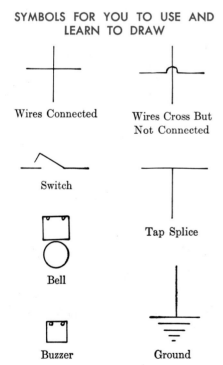

Copy the symbol drawings to help you remember them.

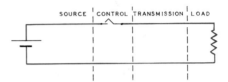

Fig. 48. Four basic parts of a circuit.

SWITCHES FOR LOW-VOLTAGE CIRCUITS

A switch is a device to open or close a circuit or, in simple language, a convenient device to put wires together or to take them apart. Conductors are fastened to insulators so that the current may be made to flow or to stop the flow. The simplest type of switch is called the *single-pole, single-throw battery switch*. See Fig. 47, left. Notice the contact points, the insulation, and the arm conductor that carries the current through when closed. Trace the circuit with a pencil, Fig. 48.

The *single-pole, double-throw switch* is used to complete one of two circuits or to break both. Trace the

THE FLOW OF ELECTRICITY

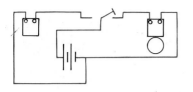

Fig. 49. A circuit using a single-pole, double-throw knife switch.

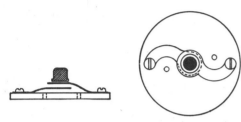

Fig. 50. The push-button switch.

paths of the current in the circuit, Fig. 49.

Remember in tracing a current: *Before electricity will flow it must have a complete circuit.* A circuit is a complete path away from the source and back to the source!

The pushbutton is another type of switch. This type of switch has two metal springs for conductors. When the pushbutton is pressed, the upper spring comes in contact with the lower spring, making contact or connection complete. See Fig. 50.

THE ELECTRIC BELL AND BUZZER

There are two types of bells, the one which rings or buzzes continuously, as shown in Fig. 51 (a buzzer is a doorbell without the bell), and the single-stroke bell. The latter is used, for example, for signaling in large buildings where a code signal is given workers who move about the building. When such a worker hears his number being tapped out on one of the many bells in the circuit in the building, he calls the telephone operator on the nearest telephone to see what is wanted.

You learned in Chapter II that a magnet may be made by running an electric current through a coil of insulated wire wound about an iron-rod core. You know that if the iron core is mild steel the core will lose its magnetism as soon as the current is turned off, but if it is high-carbon steel the core will remain a magnet longer. There are one or two of such electromagnets with soft iron cores in the electric bell.

Single-Stroke Bell

The single-stroke bell does not vibrate. The current goes in one binding post, through the coils, and directly out of the second binding post. As long as the circuit is complete (closed) from the source, the hammer is held by the magnets against the bell, and when the circuit is broken (open) the spring pulls the hammer away from the bell.

The Vibrating Bell

Follow the circuit through the diagram, Fig. 51. Starting at the right-hand binding post, the current travels up and around the top coil, then over and around the second coil to the spring that holds the armature stem and hammer, passing through the spring to the contact point and adjustment screw, and down to the other binding post. As the current makes this complete circuit through the bell,

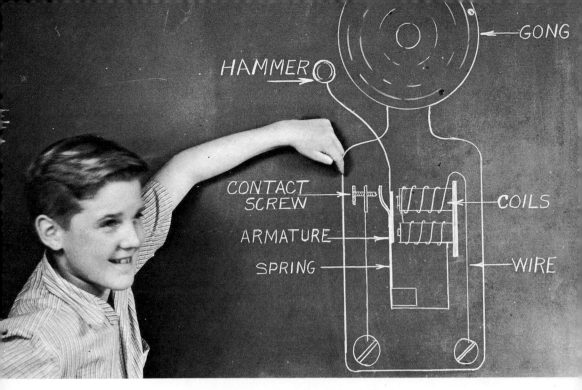

Fig. 51. The vibrating electric doorbell.

the coils become an electromagnet. The magnets attract the armature. The armature, being fastened to the stem, causes the hammer to hit the gong. As the armature is pulled to the magnets, the contact spring leaves the contact point, breaking the circuit or flow of current. When the current stops, the cores lose their magnetism and are no longer able to hold or pull the armature. A spring then pulls the armature or stem back into its normal position. Contact is made again at the contact screw, the current starts to flow, and the operation is repeated over and over again. Such action gives a bell of this type the name, vibrating bell.

TERMINALS ON METAL

Since most electrical appliances are made of metal, their metal frames are used to conduct electricity as part of the electric circuit within the apparatus. For example the base frame of an electric doorbell, being metal, conducts electricity from the "grounded" terminal to one end of the coil magnets. The frame is said to be "grounded" when it replaces one side of a circuit, thus saving wire and time.

When a frame is grounded and used as one lead, the second lead of the circuit must be insulated from the frame to avoid short-circuiting the current. Fig. 52 shows a common method of insulating a binding post from a grounded frame. Post "B" is insulated by two large fiber washers and a third one with a smaller diameter than the other two but as thick as the thickness of the metal frame. See "X". Notice that the hole

THE FLOW OF ELECTRICITY

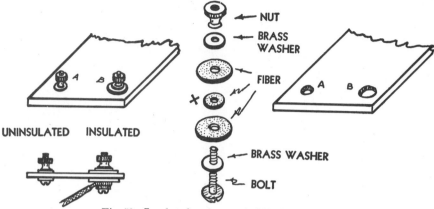

Fig. 52. Insulated and grounded binding posts.

in the frame for binding post "B" is larger than for "A". Hole "A" fits the bolt for "A". The hole "B" fits the washer "X". Examine the terminals on a doorbell.

USING A DRY CELL OR SMALL TRANSFORMER

Dry cells are widely used in low-voltage signal systems. They are designed only for occasional duty or for light loads for short periods of time, such as in bell-ringing circuits. When using them, never "short" the terminals or leave the current closed for more than a few seconds, to prevent unnecessary wear on the battery. Always screw the terminal nuts down tightly with your fingers, whether the cells are in use or not, so that the nuts are not lost.

Your instructor may prefer that you use a small, bell-ringing transformer in place of a dry cell, to save batteries. If so, make sure that the attachment cord and cap are in good working order. *Make sure that you know which two terminals you are to use to obtain the proper voltage.*

JOB 9. WIRING A SIMPLE DOORBELL CIRCUIT

Purpose: To learn how to do wiring and to make proper connections on dry cells or transformers, push buttons, and doorbells.

Steps:

1. Copy the diagram in Fig. 40 but *use the proper symbols.* Have your instructor inspect your diagram before attempting to do the wiring.

2. Obtain a panel (see Fig. 39); a transformer or cell; a pushbutton; a bell; about 3 feet of bell (annunciator) wire; a knife; a screw driver; proper screws; and a pair of pliers.

3. Fasten the bell and transformer to the panel. The button will be fastened last.

4. Allowing for right-angle turns in the wire, measure off one wire and cut to length. Be sure to allow enough wire for connections.

5. Remove 1 inch of insulation from the ends and form hooks to go clockwise around the contact screws. Loosen the contact screws but do not remove.

Working with conductors.

6. Complete the rest of the wiring.

7. Now fasten the pushbutton in place and test.

8. Have your instructor inspect and mark.

9. Remove the apparatus from the panel, straighten the wire, and return all to the instructor.

EXTRA JOB A: Replace the button in the above job with a single-pole, single-throw switch.

EXTRA JOB B: Obtain a single pole, double-throw switch, a buzzer, and more wire. Now turn back to Fig. 49 and wire the switch to ring either the bell or the buzzer. Obtain credit and return all equipment.

QUESTIONS

1. Why is salt water a better conductor than tap water?
2. What is "electronics"?
3. Which is larger, No. 18 wire or No. 14?
4. Why are there different kinds of insulation for electric wires?
5. What is flexible cord and where is it used?
6. What is a theory?
7. What are the cautions in using a knife to remove insulation?
8. What is a closed circuit?
9. Explain how a doorbell operates.
10. Why should a wire go clockwise around a binding-post screw?

LEARN TO SPELL

conductor	switch	insulator	arcing
electron	transformer	binding post	electronics
splice	proton	vibrating	buzzer

Chapter VI

LOW-VOLTAGE CIRCUIT WIRING of Signal Devices

Familiar types of circuits—bell-wiring problems—simple signal alarms and controls.

THE STUDY of circuits is best learned with safe, low-voltage current from a dry cell or a bell-ringing transformer. Connecting one bell with one pushbutton you found to be a very simple problem, but many are more difficult to understand.

Since you cannot see the electricity flow in circuits, you must "reason" where the current flows. To assist you in such reasoning, make drawings or plans first. Then if it appears to work "on paper," do the wiring.

There are many kinds of circuits, such as open circuits, closed circuits, short circuits, parallel circuits, series circuits.

The circuit is said to be open when a switch is at rest or not being pressed or closed. A broken wire, a loose connection, or a poor splice is also an example of open circuits.

A closed circuit is simply a complete circuit with the entire circuit operating. Pressing a pushbutton or making proper connections refers to closing circuits.

The short circuit, seldom desirable, refers to conditions in the circuit which are the causes of fires, "blown" fuses, etc. Short circuits often result in damage, because the "short" allows the current to flow where it was not intended to flow and without control as to the amount of current the conductor or equipment can safely handle. Fuses are used as a safety device to stop the flow if it should become too great, either through a "short" or for other reasons to be discussed later.

PARALLEL CIRCUITS

Vibrating bells are usually connected in parallel for best results. As you can see, Fig. 53, each bell obtains its current directly from the feed wire and returns indirectly to the return wire. Several switches in the same circuit require a parallel circuit also. Study the illustrations, Figs. 54, 55, showing several bells wired in parallel and several switches in parallel.

In the parallel circuit, Fig. 54, the

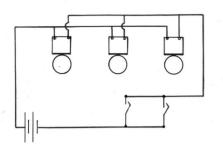

Fig. 53. Three bells controlled by either of two pushbuttons.

Notice the names of the main wires. The control wire connects the source to the *control switch;* the feed wire *feeds* to the apparatus; and the return wire *returns* the current to the source. The control and feed wires may be drawn on your plans with a red pencil and the return wire with a blue one.

The parallel circuit is also used where several buttons are desired, any one of which is to ring one or more

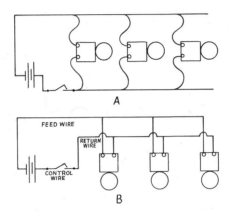

Fig. 54. Bells wired in parallel. The current is easily traced in A. The proper diagram is B.

current flows out of the negative pole through the feed wire to the bells. When the button is pushed, the current flows along the upper feed wire and to one terminal of each bell. It then flows through the bells, out of the lower terminals to the return wire, and back to the battery. All bells ring only when the current returns to the cells.

The most common way to wire bells in parallel is to place the long feed and return wires together. This circuit, Fig. 54B, is identical to the former, Fig. 54A, except that the main wires are alongside each other.

bells. See Fig. 56. Trace the current through each button.

SERIES CIRCUITS

Bells may be connected in series. See Fig. 57. But if an open circuit develops in one bell, all the rest will go "dead," because the current cannot travel beyond the open place. Therefore, unless the job specifies "series circuit," do all your wiring of apparatus in parallel.

To help you solve low-voltage circuit work with bells, learn the following three rules:

1. Connect a wire to one binding

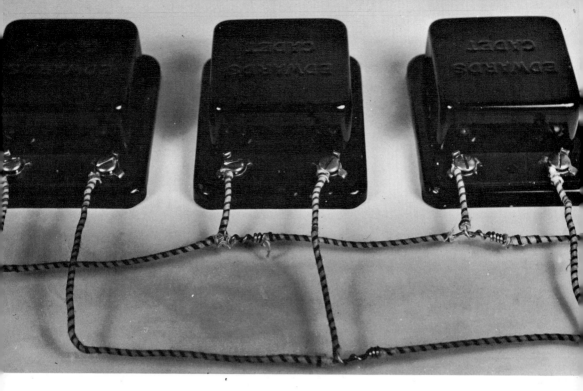

Fig. 55. Buzzers in parallel.

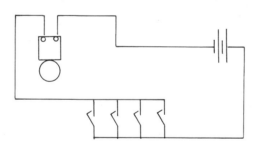

Fig. 56. Buttons in parallel.

Fig. 57. Buzzers in series.

INDUSTRIAL-ARTS ELECTRICITY

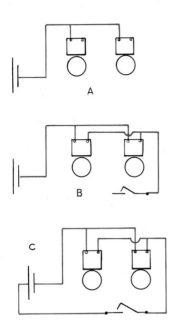

Fig. 58. Steps in parallel wiring.

post of *all bells or buzzers* and to one side of the battery. See Fig. 58A.

2. Connect a wire from one side of each switch to the other side of the battery. See Fig. 58 B.

3. Connect a wire from the remaining binding post on all bells or buzzers to one side of the switch or switches that control the bell or buzzer. See Fig. 58 C.

BELL-WIRING PROBLEMS

The many different uses for signaling with bells or buzzers in homes and in other buildings challenge the electricians to do the wiring with the least amount of wires, of other supplies, and of their time. The following jobs are typical of requests for varied circuits. The first step for the electrician is to make a drawing of his circuits, so that he knows beforehand that the apparatus will work when he has the job completed. Your first step, also, will be to make a diagram showing the apparatus in symbols and the wires. The drawing should then be checked by your instructor before you begin to do the wiring.

Since it is impossible to have you wire doorbells and buzzers in actual homes and offices, it will be necessary for you to do your actual wiring on panels or on some other type of "mock-up." Simple chimes are wired the same as bells, but chimes are made differently from bells, as already shown earlier in the book, and will be studied later.

JOB 10. WORK ORDER: The owner of a home wants the back door pushbutton to ring two bells, one in the kitchen and one in the basement.

Location Plan: See Fig. 59.

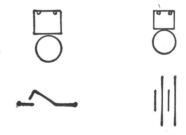

Fig. 59.

Steps:

1. Make your drawing and have it approved by the teacher. Use a red pencil line for drawing the feed and control wires and a blue pencil line for return wires.

LOW-VOLTAGE CIRCUIT WIRING

2. Obtain the materials and tools, with the exception of the wire.
3. Fasten the bells to the panel. Fasten the button with but one screw in halfway.
4. Measure the wire needed and obtain.
5. Do the wiring, following your diagram. Make all taps and splices in the approved manner.
6. Keep your wires straight and neat.
7. Have your instructor approve the wiring before testing.
8. Obtain inspection credit.

JOB 11. WORK ORDER: A screened porch is built onto an old house. Visitors now are required to enter and cross the porch to ring the chimes. The renter orders another pushbutton installed outside the porch door, to be used in the summer.

Location Plan: See Fig. 60.
Steps:

1. Make your wiring plan and have it approved.
2. Obtain materials and tools, except for the wire.

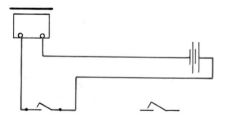

Fig. 60. Location plan for porch button.

Fig. 61. Adding a second bell circuit.

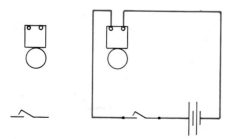

3. Install equipment.
4. Measure wire needed and obtain.
5. Do the wiring, test, and have approved.
6. Return equipment.

JOB 12. A man buys a large old house and remodels the rooms upstairs into full living quarters, so that two families can live in the house.
WORK ORDER: The man needs another doorbell and button installed in the duplex but without adding another battery or transformer.

Location Plan: See Fig. 61.
Steps:

1. Make the plan.
2. Obtain tools and supplies.
3. Wire and have inspected.
4. Return tools and equipment.

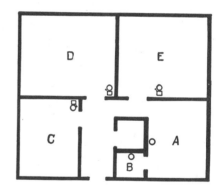

Fig. 62. School bell circuit.

Fig. 63. Circuit for the school bell circuit.

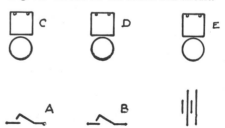

JOB 13. In a small four-room school, the principal teaches in one of the rooms and three other teachers instruct in the other rooms. WORK ORDER: The principal asks that a class signal bell be installed in the three rooms "C", "D", and "E", Fig. 62, and that two pushbuttons be so wired that either one will ring all three bells. One button is to be in his office "B" and one in his classroom "A".

Wiring Plan: See Figs. 62 and 63.

Steps: Same as in all previous jobs. CAUTION: Do not wire the bells in series!

JOB 14. A new small home is being built. The contract for the electrical work includes the installation of one pushbutton at the front door, one at the back door, and a bell and buzzer together in a central hall. The source of electricity is to be a transformer located in the basement. See Fig. 64. WORK ORDER:

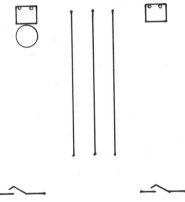

Fig. 64. Two independent circuits.

Wire the signal system for the contractor. Use only three wires to the signal units, as shown. One wire of the three is used by each unit.

Plan: See Fig. 64.

Steps: Same as all previous jobs in this chapter.

JOB 15. A business manager has an inner office for himself and an outer office for his secretary. She has a telephone on her desk with an extension to his desk. She answers all the telephone calls, and it is necessary for the manager to take only a few of these calls.

The manager requests that a buzzer be installed at "D", with a button at his secretary's desk "A". By this installation, his secretary may signal him to answer the phone when she cannot give the information desired by the person calling in.

In addition, the manager wants a button at his desk "C" connected to a buzzer in his secretary's office "B", so that he may signal for her to come into his office and take dictation.

Work Order: Wire a buzzer in each office and a button at each desk. "A" is to ring "D" and "C" is to ring "B". See plan. Only three wires are necessary between the two offices; this will reduce the cost of installation. Only one battery is necessary.

Location Plan: See Figs. 65 and 66.

Steps: Same as before.

LOW-VOLTAGE CIRCUIT WIRING

SIMPLE SIGNAL ALARMS

There are two main types of burglar alarms in common use, the open circuit and the closed circuit. The latter uses a device called a relay and will be studied later.

The simple alarm system consists of an alarm bell, a control switch to "set" or "shut off" the alarm, and an alarm switch. Fig. 67a shows such an installation. Note that the control switch, the bell, and the alarm switch are connected in series. If the control switch is closed, the bell will ring only when someone opens the window and closes the circuit unknowingly. Fig. 67b shows a door alarm switch used commonly in small shops to tell a worker in a back room that a customer has entered the front door to the store.

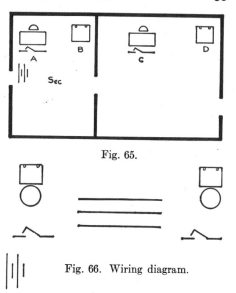

Fig. 65.

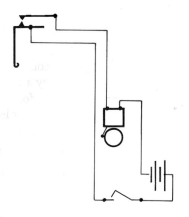

Fig. 66. Wiring diagram.

Fig. 67a. Alarm switches, one installed on a model door.

Fig. 67b. Alarm switch circuit.

EXTRA JOB *A*. We have here a four-family apartment house. Install a system whereby a person wishing to enter the building can signal an apartment by means of a bell within the apartment, which is controlled by a pushbutton at the front entrance, numbered the same as the apartment. If the people in the apartment find, after talking

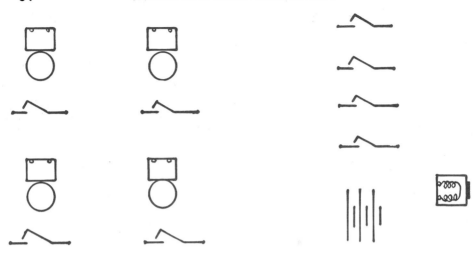

Fig. 68. Apartment signal circuit and lock control.

with the person at the front door, over the phone or through the intercom system, that they want to let the person in, they press the button in the apartment which controls the lock on the front door; this allows the person at the door to open it from the outside. Wire this system on your board.

EXTRA JOB *B*. Obtain a window or door, open-circuit, alarm switch, and wire a signal or burglar-alarm circuit.

EXTRA JOB *C*. Figure 69 shows typical wiring of a bell installation within the walls of a home. Make a drawing of this, adding a wall on the left and showing the back-door button, the buzzer on opposite side of wall from the bell, and the connections to the transformer. The drawing paper should be at least 12″ x 15″.

In Fig. 69, holes are drilled through the 2″ x 4″ plate and the rough flooring at "A", one through the lath and plaster to the bell at "B", one through the floor plate and the rough flooring at "D", and one through the sheathing, the paper, and the outside finish to the button at "E". The wires are hidden in the walls at "B" and "E". On a new house, wiring is done before plastering; appliances are installed afterwards. Trace the circuit.

EXTRA JOB *D*. The make-and-break push button has two contacts. One circuit is always closed until the button is pushed; this contact closes the second circuit and opens the first. Obtain two such push buttons, two buzzers, two dry cells, and sufficient wire to connect a return-call, two-wire circuit. Show that pushing "A" buzzes "Y", and that "B" buzzes "X". Show the difference between this installation and the one in Job 15.

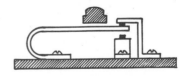

Cross section of a make-and-break push-button.

LOW-VOLTAGE CIRCUIT WIRING

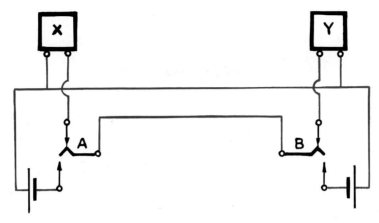

A circuit for the make-and-break pushbutton.

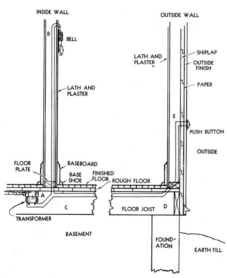

Fig. 69. Wiring within the walls.

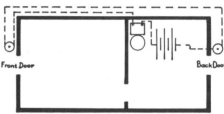

2. Suppose you connected a coil about a nail to a battery to make a magnet. The coil does not operate because the wire is broken inside the coil. Is the place where the wire is broken called a short circuit or an open circuit?

3. A nail is placed accidentally across the two terminals of an electric bell, and it does not ring. Is this called a short circuit or an open circuit?

4. Why do bells "flutter" when connected in series?

5. Why is it necessary in wiring to include the drawing or plan of the connections on all circuit-wiring jobs before you start the job?

QUESTIONS

1. What is wrong with the wiring in this drawing?

LEARN TO SPELL

circuit	inspection	wire gauge	safety
pliers	apparatus	Western Union	dry cell
asbestos	signal	adjusting screw	pushbutton

Chapter VII

HEAT FROM ELECTRICITY Applied to Everyday Problems

Meaning of "resistance"—resistance measured in Ohms—resistance and voltage—the rheostat—Ohm's law—heating appliances—fuses—other applications of heat from electricity.

ELECTRICITY that comes into the home may be used to produce heat for cooking, for toasting bread, for heating air, or for freezing water or food. It seems strange that electricity can be used for either heating or freezing!

One could make a fairly long list of electric appliances commonly used in the home to produce heat, such as the toaster, the flat iron, the mangle, the heating pad, the bottle warmer, the portable heater, the electric stove, the percolator, the waffle iron, and the water heater. Can you add any other appliances to this list?

A small child was once playing on the floor and noticed a group of small holes in the wall near the floor. A hairpin lay on the rug near her, and she decided to investigate what was in the little holes. She pushed each point of the hairpin into a hole of the electrical outlet. The electricity that was conducted through the hairpin so paralyzed her that she couldn't let go of the hairpin until her mother, hearing her scream, pulled her away. In that short time, however, the hairpin had turned red hot, burning deeply into the flesh of her hand.

The little girl found (but not too happily) that electricity produces heat! Why did that wire get hot? The answer is that the heat was caused by the *resistance* in the piece of wire.

RESISTANCE MEASURED IN OHMS

We have noted the varied qualities of conductors and found that copper is a good conductor of electricity, that iron is fair, and that nichrome is poor. Although every metal has a tendency to resist or to hold back the current to some extent, copper hinders the flow of electricity less than the metal nichrome does. To use the proper word, copper "resists" the flow of electricity less than nichrome "resists" its flow. Or, the other way

HEAT FROM ELECTRICITY

around, nichrome offers greater resistance to the flow of an electrical current than does copper.

A boy's sled will "flow" or run easily over shiny ice; but concrete pavement upon which there is no snow or ice offers a great deal of resistance to the "flow" of the sled.

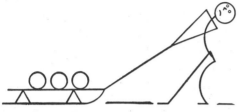

When electricity is made to run from a source through wires and switches and joints and various appliances or electrical devices and then back to the source, the pressure naturally is slowed down by the resistance to the flow offered by all these various conductors. Now, resistance can be measured, and the term describing the extent or amount of resistance is called the ohm.

The length of wire, the size (gauge), and the material of which it is made are all concerned with the resistance of conductors.

For example, suppose you were to take two pieces of wire, one of copper and one of nichrome. Each wire is 20 feet long and of 24-gauge (the diameter). If each were connected to two separate 110-volt outlets, the copper wire, being a good conductor, would probably short the circuit and burn out the fuse; and the nichrome wire, being a poor conductor, would get red hot its full length.

The following is a table showing the diameters of various wires and the resistance in ohms, along with other information about solid copper wire and nichrome wire. Compare the resistance of 24-gauge copper wire with that of 24-gauge nichrome wire. You find that the ohms resistance of 100 feet of copper wire is 2.567 ohms: and the resistance of 100 feet of nichrome wire is 1.62 ohms x 100 or 162 ohms. Compare the two figures of 2.567 ohms and 162 ohms. Nichrome has a resistance of nearly 64 times that of copper!

WIRE TABLE

		SOLID COPPER WIRE				NICHROME WIRE		
Gauge	Diameter	Amps. with safety	Line drop per 100 ft. Ohms	Lbs. per 100 ft.	Ohms per 100 ft.	Length	Watts	Ohms per foot
8	.1285	35	2.2	4.998	.0628	x	x	x
10	.1019	25	2.5	3.142	.0998	x	x	x
14	.0641	15	3.79	1.243	.2525	x	x	x
16	.0508	6	2.41	.7818	.4016	x	x	x
18	.0403	3	2.02	.4917	.6385	29'2"	1000	.406
20	.032	x	x	.3092	1.015	28'5"	650	.634
22	.0253	x	x	.1945	1.614	25'7"	450	1.01
24	.0201	x	x	.1223	2.567	24'10"	300	1.62
26	.0159	x	x	.0769	4.081	23'6"	200	2.57
30	.0100	x	x	.0304	10.32	24'7"	75	6.5
32	.0079	x	x	.0191	16.41	23'7"	50	10.15

INDUSTRIAL-ARTS ELECTRICITY

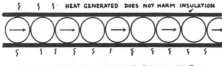

THIS WIRE IS JUST THE RIGHT SIZE—
HEAT GENERATED IS SAFELY DISSIPATED

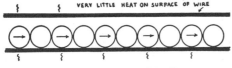

WIRE BEING OVERSIZE,
VERY LITTLE HEAT IS GENERATED

HEAT GENERATED MAY HARM
INSULATION OR MELT WIRE!

COURTESY, WESTINGHOUSE

Fig. 70. Crowded electrons produce heat.

But why did the nichrome wire get hot while the copper wire did not? The nichrome wire as a conductor for electricity may be compared to the concrete as a conductor for the boy's sled. If a heavily loaded sled were pulled by a strong force over concrete pavement, wouldn't the runners of the sled and the concrete where the runners dragged get a little warm?

The free electrons of the atoms are forced through the nichrome, which offers a great deal of resistance to the flow. See Fig. 70. The electrons, being crowded and inclined to "pile up," produce heat and make the nichrome wire hot.

RESISTANCE AND VOLTAGE

Through any given conductor of electricity, the resistance of the conductor affects the balance between the amount of electricity or amperage flowing, and the pressure of voltage forcing it along. Suppose a boy is pulling a rock on a sled on a dry concrete sidewalk. The boy has only so much force to exert. If you add more rocks, one at a time as he pulls the sled along, there will come a time when one additional rock will "stall" the sled. To move the stalled sled still farther, he will require help from someone else who adds *his* force, or voltage. Now the two boys pull the heavily loaded sled over the concrete sidewalk and then onto an icy stretch where only one boy is needed to pull the sled.

The three factors that varied in the above illustration were:

1. Resistance of the conductor (in this case concrete and ice)

2. Amperage (in this case stones)

3. Voltage (in this case pull power)

THE RHEOSTAT

A conductor such as nichrome wire or German silver wire, (often called nickle-silver wire), which offers resistance to an electric current, is valuable as part of a device to control the amount of current in a circuit. Fig. 71 shows a rheostat consisting of a coil of such wire; over it slides a movable contact arm. Trace the current and notice that, as the arm is turned clockwise, more current is required to flow through the resistance coil. The coil acts as a damper to vary the amount of current desired and to dim or brighten the light.

HEAT FROM ELECTRICITY

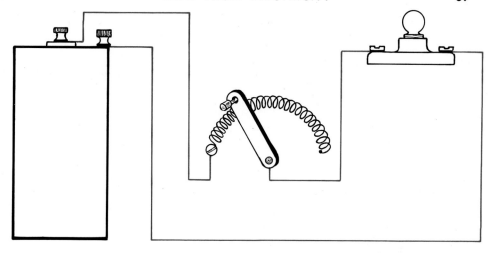

Fig. 71. The rheostat.

The rheostat is used in radios for such purposes as to control the volume; it is also used to dim or brighten stage lights and for hundreds of other purposes.

OHM'S LAW

In designing, installing, or servicing electrical apparatus, electricians often find it necessary to know the resistances in the various circuits. Then, too, if they know the resistance, it is sometimes necessary to find how much current will flow or to find the force of the current. Volts, amperes, and ohms are the three units concerned. If two of these units are known, the third may be found by using Ohm's law. Ohm's law is a statement which, simply expressed, is as follows:

The total amperage multiplied by the total resistance is equal to the voltage, or necessary force.

The following three problems illustrate the use of the Ohm's law.

$$\boxed{\text{amps}} \times \boxed{\text{ohms}} = \boxed{\text{volts}}$$

CASES

A. When voltage is not known:
$2 \times 55 =$? or
$2 \times 55 = 110$ volts
2 amps $\times$ 55 ohms $= 110$ volts

B. When the resistance is unknown:
$2 \times \boxed{\text{what no.}} = 110$
$110 \div 2 = 55$ or amps $\boxed{\dfrac{\text{ohms}}{\text{volts}}}$

C. When amperage is not known:
$\boxed{?} \times 55 = 110$ or
what number multiplied by $55 = 110$?
$110 \div 55 = 2$ or
ohms $= \dfrac{\text{volts}}{\text{amps}}$

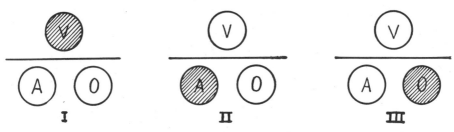

Fig. 72. Resistance formula.

In case *A*, two amperes flow in a conductor having a resistance of 55 ohms. The voltage necessary to force the two amperes is 2×55, or 110 volts.

In case *B*, two amperes flow through a circuit with a force of 110 volts. What is the resistance being offered by the conductors? Two times what number equals 110? ($2 \times 55 = 110$ or $110 \div 2 = 55$ ohms.)

In case *C*, the resistance of the circuit is 55 ohms, and the voltage is 110. How much amperage is flowing? What number multiplied by 55 equals 110? ($2 \times 55 = 110$; $110 \div 55 = 2$)

For students who have had simple algebra, Ohm's law is expressed in three equations. Here the letters "E" for volts, "I" for amperes, and "R" for ohms are used:

$$I = \frac{E}{R} \quad R = \frac{E}{I} \quad E = IR$$

As a memory aid, Fig. 72 shows letters placed in circles above and below a line representing the expression for an arithmetic fraction. To find any one of the three, cover one circle with your index finger and the answer will be apparent. For example, in "I", the "V" for volts is covered and the answer, below the line, is equal to amperes multiplied by ohms or AO. In "III" ohms is unknown and covered; the answer is equal to the volts divided by amperes.

Engineers who design electrical appliances use Ohm's law in estimating the type of materials to be used in the appliances.

HEATING APPLIANCES

Electric hot-water heaters, electric heating furnaces, and electric ranges are usually permanently installed and require expert attention when something goes wrong. Hot-water heaters and ranges have heating elements of varied design. The elements are coils of resistance wire which become warm when electricity flows through them, most of which work on the principle of electricity flowing through coils of resistance wire.

The hot-water heater is often installed in the home with an electric-clock-control unit which permits the use of electricity for water heating at the periods in the day when electricity is used the least for other purposes.

Because the electric percolater, waffle iron, flat iron, mangle, toaster, oven, and heater are portable appliances, the flexible cord attached to these appliances often becomes worn or broken and should be inspected fre-

HEAT FROM ELECTRICITY

COURTESY, WESTINGHOUSE
The electric blanket

COURTESY, GENERAL ELECTRIC
The electric iron

COURTESY, WESTINGHOUSE
The electric roaster

Fig. 73. Heating applications used in the home attached to source with flexible cord.

quently to find any evidence of possible trouble. Fig. 73.

Keep in mind that, to produce heat, a great deal of electricity is needed, especially in comparison to the amount needed for lighting. The typical flat iron, for example, uses as much electricity per hour as ten ordinary electric light bulbs.

The Care of the Electric Iron

When an electric iron does not heat, the trouble may be due to an open circuit in the cord; or the heating element inside may be burned out and need replacing; or the control device, if the iron has one, may be defective and need replacement. It is not difficult to care for the cord. Your next job will be to make an appliance cord for an electric iron.

Many accidents happen daily in the use of electric irons, most of which could be prevented if the following precautions were observed:

1. Attach the cord directly to a permanent fixed outlet, never to a swinging socket.

2. Detach the cord by pulling on the cap that protects the connection.

The electric range
COURTESY, FRIGIDAIRE, DIV. OF GENERAL MOTORS

Electric hot-water heater.
COURTESY, GENERAL ELECTRIC

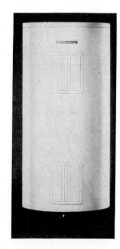

Fig. 74. Appliances installed in the home. Appliances such as stoves, water heaters, furnaces (controls), attic air conditioners, etc., must be installed by licensed electricians, as required by local ordinances.

Fig. 75. Left to right: plastic or rubber-covered extension cord; rubber-covered, reinforced two-wire cord for vacuum cleaners; asbestos-wound heater cord.

Fig. 75a. Common types of caps.

Never pull on the cord, as you may pull the wire ends from the contact screws.

3. Inspect the cord each month to see if there are any weak parts. The wire usually breaks just outside of the spring in the heating-appliance plug, because the person who irons twists the cord in using the iron. Repair or replace all defective cords.

4. Use an iron with a control element to prevent overheating.

5. Use an outlet with a red pilot light to remind the user that the iron is "on" or "off."

6. Replace broken plugs or caps.

Flexible Heating-Appliance Cord

Buy only those cords which have the tag showing approval of the Underwriters Laboratories. See Fig. 75. Approved cord consists of:

1. Twisted-strand copper wire.

2. Asbestos insulation about the wire.

3. Heavy cotton, braided cover.

4. Approval tag attached.

JOB 16. To Make an Attachment Cord for Electrical Appliance

Materials: Attachment plug cap; appliance plug; asbestos-insulated heater cord No. 18; knife; thread; screw driver; pliers; tape.

Steps:

1. Determine the length of the cord. Standard length is 6 feet but the cord may be made longer to meet requirements.

2. Slip the cap over the wire and connect as in Fig. 76, "B". The wires should go clockwise around the terminal screw.

3. Remove 1½ inch of the outside braid and wrap with thread to keep it from unraveling. Remove ¾ inch of insulation from each end of the wires. Do not cut the fine wires. Twist these fine wires with your fingers until they form a tight cable. Keep the asbestos on the wires and wrap each wire with tape ⅛ inch wide, or with thread, as in Fig. 76, "A".

4. Open or separate the appliance plug and see how and where the cord goes, so that you can prepare the end to fit in the grooves for safe insulation and snug fit.

5. Slip spring on end of cord.

6. Proceed as in *2* above.

7. Place wires in grooves and press down tight. No material should come between parts when placed together.

8. Connect the ends of the wires under terminal screws.

9. Put spring in place.

10. Put other half of the plug in place, then replace screws or clips.

11. Get your instructor's approval and test.

It is recommended that all appliances and machines, even in the home, such as washers, driers, deep freezers, refrigerators, grinders, drill presses, power saws, hand power saws, etc., be

HEAT FROM ELECTRICITY

grounded. This is especially important when they are placed in basements or garages. See the picture of the adapter plug for use in grounding a circuit into the standard outlet. Note the attachment cord cap with the grounding post. Also, note the wire from the adapter which grounds the circuit to the outlet box, through the cover plate screw. Machines not equipped with the grounded attachment caps can be easily grounded. Tape a flexible rubber-covered wire to the attachment cord, ground the cap end to the cover plate screw and the other end to a convenient grounded bolt or machine screw on the motor.

The Waffle Iron

Servicing the waffle iron, with the exception of the cord, is not an amateur's job. If either of the two griddles (top and bottom) does not heat, the trouble may be located in the one which has failed. The element within the griddle may need to be replaced. Occasionally the trouble may be in the wires in the hinge leading to the upper element. However, one should always check the appliance cord first before taking the iron to be serviced.

Treat the waffle-iron cord as you do the flat-iron cord. Never wash the iron by dipping in a pan of water. Wipe off with a clean cloth.

The grounded plug adapter.

The Electric Toaster

Most manufacturers of electric toasters include complete instructions for the care of their product. Never dig toast out with a fork or sharp object, as you may damage the heating-element wire. Treat the cord as you would any other appliance cord.

Miscellaneous Appliances

Care for the cords to the percolator, mangle, oven, curling iron, and portable heater the same as you care for the iron cord. Instructions for their care and us are usually furnished by their manufacturer.

Bottle Heaters and Vaporizers

Now in common use are a bottle warmer and a vaporizer so constructed as to shut off the current automatically when the water level is lowered by action of the current in heating or vaporizing.

Fig. 76. Wiring the attachment plug cap.

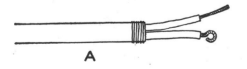

A

B

A feed-through switch should be installed in the cord, so that one is not tempted to pull out the appliance plug by pulling on the cord, which might cause tipping the boiling water over on the hand.

Read and observe the instructions given with these appliances. If they are of the type which do not open the circuit automatically when the water is boiled away, do not allow them to operate after the water has gone. Fires have been started in homes with such steam vaporizers when they were not used with care.

FUSES

A fuse is a safety device and must be put in every circuit where there is danger of overloading the line. All the current used must pass through the fuse. The fuse is the weakest link in the circuit. If a short circuit or overload were to cause more current to flow than the carrying capacity of the wire, the wire would become hot and set fire to the insulation or any inflammable matter near, were it not for the fuse. All fuses have a number stamped on them showing the carrying capacity in amperes. If the flow is greater than the carrying capacity of the fuse, the fuse melts, or "blows," resulting in an *open* circuit.

In all fuses, the metal alloy link is enclosed to prevent a spark or flash when the fuse blows. There are two general types of fuses. One is the *plug fuse*, which is the most common household type, ranging from 5 to 30 amperes and may be removed easily from the fuse block. The other is the *cartridge fuse*, which is made to fit into clip holders. Some are renewable—that is, a new fuse link may be put in after they "blow" and after the wiring error has been corrected.

Fuses are used in the auto electric circuit along with circuit breakers. (See Workbook Fig. VIII-12, p. 55.) Today most power brought into buildings enters through a circuit breaker rather than through fuses. Circuit breakers are devices that disconnect the power mechanically. They contain simple magnets in the form of solenoid coils. The current flows through the coils, which are in series with the electric load on the line controlled by the circuit breaker. If the load calls for more amperage than is safe for the wiring in the circuit, the solenoid begins to operate and breaks the circuit.

Motors require much greater amperage, when first started, than when running. The fusetron fuse is one which permits a heavy flow of current for a very short time, to prevent fuse blow-outs when a load is unusually

Fig. 77. Types of fuses and circuit breakers.

HEAT FROM ELECTRICITY

heavy at the start. Your instructor will show you such a fuse. Compare it with the old type.

When a fuse "blows," it indicates trouble. Perhaps a short has occurred; for example, insulation has worn off, so that bare wires come together inside the appliance plug on the electric flat iron. Perhaps too many appliances have been connected at one time, and an overload of current has been called upon to do the work.

Always have a flashlight handy for such power loss emergencies. First remedy the cause; then re-start the breaker or change the fuse. If it occurs again, you may need a repairman. A burned out fuse is detected through the mica window. Do not replace with one of higher amperage.

No. 14 rubber-covered wire is the wire commonly used in the home. Number 14 wire will safely carry 15 amperes. Therefore a 15 amp. fuse should be the largest that one should put in to insure maximum safety. If the fuse box is located in the basement or in a damp place, pull the main switch before attempting to change fuses. Don't work in the dark, but use a flashlight or candle. To eliminate your fuse protection by putting a coin behind the fuse, rather than a complete replacement, is foolish and inviting disaster.

OTHER APPLICATIONS OF HEAT FROM ELECTRICITY

In a previous experiment with the induction coil, you were asked to observe a spark as it jumped across an air gap between two carbons attached to the secondary circuit or the coil.

This arc is the effect of electricity as it jumps through the air. What you see is not electricity but the result of the electric force producing heat and burning the carbon in the open air. Electric furnaces for melting metals which work on this principle are used in industry, but, of course, the coils are larger and a great deal more current is necessary.

Arc Welding

Electric welding is a process of joining two or more pieces of metal by slowly moving the tip of a special welding rod close to the pieces to be joined while a current is passing through the rod and through the pieces to be welded. An electric arc is "struck" or generated at the point of contact between the pieces and the rod tip. The heat from the arc melts the rod slowly, as well as the metal at the place being welded; and the pieces to be joined are fused together.

Suppose a steel frying pan or skillet develops a crack across the bottom, and the owner wishes to have it welded. The welder places the pan upon a metal-covered table and fastens one contact from the welding machine to the table, which is metal and therefore a conductor of electricity. He protects his eyes with a helmet, fastens a welding rod in a special holder attached to the second contact, and turns on the current. He "strikes an arc" at the end of the crack with the rod tip and begins to slowly "run a bead" down the length of the crack.

Another method of producing heat from electricity is by changing the usual current supply with the transformers and other apparatus to short radio waves. Even the principle of short radio waves is too far advanced to explain here. In the practice of

SYMBOLS

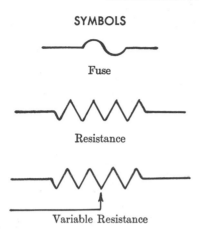

Fuse

Resistance

Variable Resistance

medicine this type of heat is used to penetrate far below the surface of the body and thus apply heat to an injury which cannot be reached by heat from a hot-water bottle or heating pad. The process is called diathermy.

Still another method of using electricity to produce heat is by use of a device called a "high-frequency heating unit." Again, the process is too involved to explain in this text, but the principle is similar to that used in diathermy.

EXTRA JOB *A:* Obtain an old electric flat iron, waffle iron, curling iron, or toaster. Using a cigar box for storing the screws and parts, disassemble. Make a cross section sketch marking various parts to assist you in remembering how to put it back together. Have your instructor check the disassembly. Reassemble and test.

If any of the screws or bolts are rusted, soak them with light oil or kerosene before attempting to unscrew. Should a screw or bolt break off, see Extra Job *B*.

EXTRA JOB *B:* Remove a broken-off bolt by drilling a hole through the center of the bolt and turning the bolt out with an "Ezy Out."

EXTRA JOB *C:* Inspect, repair, or

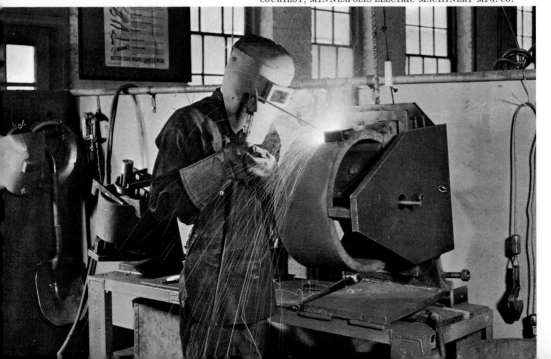

Arc welding a generator frame

COURTESY, MINNEAPOLIS ELECTRIC MACHINERY MFG. CO.

The home air conditioning control.

Air conditioning in today's home includes heating, cooling, and humidity and air pollution control. One approach to the ideal indoor weather is the Weather Control Center, a master control panel, to regulate the heating, or cooling, of each room in the house.

replace heating-appliance cords on appliances brought from home.

EXTRA JOB D: Make a collection of different types of fuses or circuit breakers. Arrange and mount on a panel. See if your instructor will allow you time to describe your collection to the class.

QUESTIONS

1. An electric iron has a resistance of 20 ohms and requires 110 volts to operate. How many amperes does the iron draw?

2. Suppose that there are other appliances and lights which take a total of 8 amperes more than the amount used in the problem above. What ampere fuse should be used in the circuit?

3. What is the resistance of 50 feet of No. 10 copper wire? See Table, page 67.

4. How many feet of No. 22 nichrome wire would be needed to make a toaster with 1.01 ohms resistance per foot? See Table, page 67.

5. Is No. 14 wire larger in diameter than 24-gauge?

6. What does 50 feet of No. 18 copper wire weigh?

7. If No. 18 bell wire costs 60 cents per pound, what will 100 feet cost?

8. Examine the insulation in a toaster or flat iron and trace the circuit.

9. Why is flexible wire used in appliance cords?

10. Why is it not "good sense" to use a penny behind a burned-out fuse?

LEARN TO SPELL

resistance	ohm
nichrome	rheostat
flexible	fuse
cartridge	overload
mica	frequency
appliance	element

Chapter VIII

LIGHTING WITH ELECTRICITY and Electrical Lighting Devices

The electric-light bulb—portable lighting—low-voltage lighting circuits—two-way switch for lights—lights in parallel and series—meter reading—extension cords—sockets—feed-through switches and other problems.

Most people who live in homes or work in buildings which are electrically lighted become so accustomed to the service that they seldom, if ever, stop to appreciate the convenience or bother trying to understand why the bulbs light.

As used for lighting, electricity serves man well. Streets are lighted so that he may walk safely at night. Portable lights such as flashlights help him to find things in dark places. Doctors can examine the patient's throats, and even their bronchi, by the use of very small lights. Electric lights are placed on high buildings to guide aviators away from danger and on towers miles apart to help guide planes to their destinations. Thousands of uses are made of electricity for lighting, so that the eyes of man may see in the darkness.

THE ELECTRIC-LIGHT BULB

You have just learned how electricity flows through a resistant wire to produce heat. When the wire gets hot, it also glows and produces some light. Edison searched a long time until he found a bulb filament which would produce light but very little heat. He placed the wires inside a glass bulb from which he removed most of the air, because the oxygen in the air united with the material in the wire to burn up the wires. His first successful light bulb, however, did get extremely hot. Later, discov-

Fig. 78. Electrons flowing through a light-bulb filament.

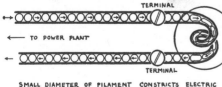

SMALL DIAMETER OF FILAMENT CONSTRICTS ELECTRIC CURRENT, CAUSING MUCH FRICTION. THIS FRICTION HEATS UP WIRE TO WHITE HEAT (AND NO MORE)

COURTESY, WESTINGHOUSE

LIGHTING WITH ELECTRICITY

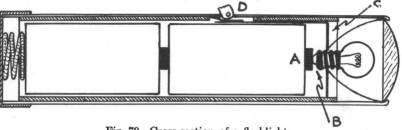

Fig. 79. Cross-section of a flashlight.

eries by other men led to the development of light bulbs which give a whiter light than Edison's and with much less heat.

Suppose we connect two copper wires to light a bulb. The current flows freely through the copper wire leading to the filament wires. The filament in almost all light bulbs, as noted in Chapter VII, is made of tungsten wire, because the metal tungsten has high resistance and a *high melting point*. The electrons of the tungsten atoms have a difficult time making their way through this high-resistance filament. The friction resulting from the electrons being forced through such material results in the material's becoming white hot and giving off light. When the electrons pass through the tungsten to the copper return wire, they again have ample area to go merrily on their way. See Fig. 78.

PORTABLE LIGHTING

Of course, a light bulb must have a source of electricity. The common flashlight has one or more dry cells as the electrical source for its bulb. The flashlight dry cell is but a smaller type of the telephone cell described in a previous chapter except that the connections are made without the use of wires and the screw binding posts.

Trace the circuit through the flashlight shown in Fig. 79. The spring holds the cells tightly in the case and insures good contacts. The current travels through this spring as it starts from the negative or − end of the cell and continues to the switch "D" and along the conductor "C" to the side of the light bulb, through the filament of the lamp and out the center disk "B" to the positive or + pole of the cell. Cells in the flash light are in series.

There are many kinds of fully portable electric lights, ranging in size from the small single-cell light attached to a key case and used to light up the keyhole to the long, five-cell flashlight used as a spotlight. Many railroad brakemen use a lantern-type flashlight. Flashlight batteries wear out rapidly. This is the one main objection to portable lights.

LOW-VOLTAGE LIGHTING CIRCUITS

Electricity for most houses and other buildings is received from power stations through wires connected to near-by power lines. These power stations are often many miles away.

Since the electricity coming into your home or school is dangerous if not used properly, you will first learn

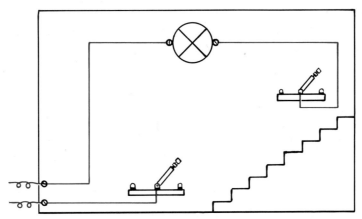

Fig. 80. The stairway circuit.

to use current from a low-voltage source such as a dry cell or a transformer. Miniature receptacles (sockets), bell wire, and knife switches are some of the material you will use.

JOB 17. SIMPLE LIGHTING CIRCUIT

Purpose: To wire three miniature light receptacles to be controlled by one switch.

Steps: (Read this entire procedure before following it.)

1. Obtain three miniature porcelain base receptacles; 4 feet of bell wire; screws, screw driver; one dry cell or transformer; and one battery switch. Make a wiring diagram of the circuit and have it checked by the instructor.

2. Fasten three miniature receptacles to a wiring panel; arrange in a row about an inch apart.

3. Install a battery switch or a single-pole, single-throw knife switch on the panel. See Fig. 47.

4. Connect the three sockets in series to the switch. Obtain and install three Christmas-tree light bulbs or three 1½-volt flashlight bulbs.

5. Connect the lead terminals to a 15-volt transformer or, if flashlight bulbs are used, to a dry cell.

6. Notice how dimly the bulbs light.

7. Now connect the three sockets in parallel, to be controlled by the switch.

8. Turn on the *same* voltage as used when the lights were wired in series. Now note how brightly the bulbs light.

THE TWO-WAY SWITCH FOR LIGHTS

One of the most interesting lighting circuits is that which requires the control of the lights by either one of two switches, such as the circuit used in wiring stairway or hall lights. Your next job will represent such a circuit, using two single-pole, double-throw knife switches to control the lights from either of two switches. See Fig. 80. If you understand how the current travels in this circuit, it will help you to understand the two-way and three-way snap switch and toggle switch

LIGHTING WITH ELECTRICITY

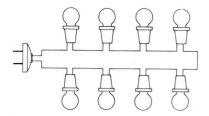

Fig. 81. Lights in series.

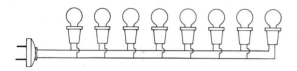

Fig. 82. Lights in parallel.

used later in stairway or hallway circuits. The stairway light must be so wired that it may be turned on or off either at the foot of the stairs or at the top of the stairs.

JOB 18: LIGHT WIRING

Purpose: To wire a light so that it may be controlled by either of two switches.

Steps:

1. Obtain two single-pole, double-throw switches; one miniature socket; 5 feet of bell wire; one bulb, a dry cell or transformer; and necessary tools and screws.

2. Make a wiring diagram by copying and completing the one below, Fig. 80, and have it checked.

3. Wire the circuit on a panel and test.

4. Each switch must be closed on one side or the other.

LIGHTS IN PARALLEL AND SERIES

The common single-wire Christmas-tree light string has eight lights (Fig. 81), the bulbs of which are connected in *series*. Each bulb takes 15 volts pressure. Eight bulbs of 15 volts each in series take 8 times 15 volts, or 120 volts for the entire circuit. These strings of lights are made especially for use in homes wired for 115 volts.

If there were but six 15-volt bulbs in a series-wired string instead of eight, the bulbs could take a voltage of 6×15 volts, or 90 volts. Then, if the six lights were connected to a 110-volt outlet, all the lights would burn out because they could not stand the total pressure of the 110-volt current. On the other hand, if ten 15-volt bulbs were connected in series to a 110-volt outlet, they would light but very dimly.

The objection to a series circuit is that if one light burns out all the lights go out. Also, the right number of bulbs is always required to match the voltage. Too few, and they burn out. Too many, and they are too dim.

A bicycle tire is made to hold twenty pounds of air pressure. This is a similar case to series-wired lighting. If forty pounds of air were put into the tires, the casing would blow out; and if only ten pounds of pressure were put in, the tire would be soft.

The lights in the more modern Christmas-tree string (Fig. 82) are wired in parallel. Each bulb requires 115 volts of pressure in order to burn brightly. In a parallel circuit of lights, if one light burns out, all the other lights stay burning because *each* bulb is provided with the necessary **115** volts.

All general house wiring for lighting is done in parallel, so that each outlet may be controlled without interfering with any other part of the wiring system. See Fig. 83 and trace the circuits throughout the various rooms in the house.

Fig. IX-1, Workbook p. 59, shows other details of the same house wiring arrangement shown in Fig. 83. Study these two drawings. Note that three wires are attached to the meter from the pole. Half the outlets are connected to one outside wire and the center wire. The other half are connected to the other outside wire and the center. Two separate circuits, then, run to the different rooms to balance the load. Knife switches and fuses are shown in the wiring diagram to clarify the job of the circuit breaker. Also shown are the steel conduit and outlet boxes which contain the wires of the circuit.

Fig. 83, then, is a wiring diagram also called a schematic. The drawing on page 59 of the Workbook shows how it is actually installed.

WATTS

Electricity is not a material like water or air, because either water or air is a *substance;* and electricity is a *force*. Wind is air in motion. A gale is air in a hurry! Electricity may be thought of as more like wind than air because wind expresses the *force* of the air, rather than air itself. Wind has been known to force a straw into a tree. Wind can be dangerous if it blows with much force. So may electricity be dangerous if the voltage and amperage are great.

You have learned that the word voltage means force or pressure, and amperage means volume or amount. To prevent confusion, you were not told the whole story about amperage. Amperage means volume, true, but amperage is more than that. It is a given volume, flowing *at a certain rate*.

A new term, which is probably already familiar to you, is *watts*. Watts is the voltage of the current flowing multiplied by the amperage. $V \times A = W$. Electric-light bulbs used in homes are marked to show the voltage necessary to "light" them, and the wattage is marked to indicate the brilliancy. In our world, where the use of electricity is so common, nearly everyone knows that a 110-watt bulb gives more light than a 40-watt bulb of the same type.

If volts multiplied by amperes equal watts, it is possible to find one of these three by arithmetic or algebra if only two are known. For example, if a 110-volt bulb draws 1 ampere of current per hour, $V \times A = W$ or $110 \times 1 = 110$ watts. The bulb uses 110 watts per hour. A 100-watt bulb at 110 volts requires .909 amperes per hour. 110 multiplied by what number equals 100?

$100 \div 110 = .909$ or .909 amperes

PROOF: $110 \times .909 = 100$ watts

Find how many amperes are required to light a 110-volt bulb of a 1000-watt rating. If two of such bulbs were connected to the same circuit, protected by a 15 ampere fuse, would the fuse burn out?

LIGHTING WITH ELECTRICITY

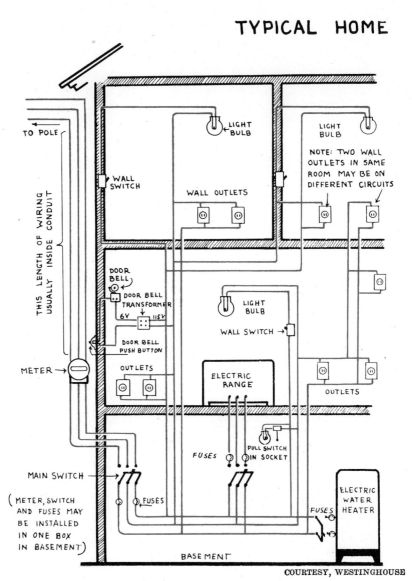

Fig. 83. Electric circuits in the home.

Fig. 84. Electric house meters.

You will use this problem of VA = W many times in the remaining part of the course. Make sure now that you know how to work such problems.

Since so many watts are used in homes or in other buildings, a larger unit, called the kilowatt, is used for convenience. *A kilowatt is 1,000 watts.* The meter is called a kilowatt-hour meter because it measures kilowatt hours. A kilowatt hour is a unit of work or energy equal to that done by one kilowatt of electricity acting for one hour.

Examine a light bulb, and find the number of watts the bulb uses. Examine the name-plate on a toaster. A toaster rated at 600 watts uses 600 watts or .6 kilowatts per hour.

$$\text{watts} = \text{amperes} \times \text{volts}$$
$$\text{amperes} = \text{watts} \div \text{volts}$$
$$\text{volts} = \text{watts} \div \text{amperes}$$
$$\text{kilowatts} = \text{amperes} \times \text{volts,}$$
$$\text{divided by 1,000}$$
$$\frac{\text{amperes} \times \text{volts} \times \text{hours}}{1,000} =$$
$$\text{kilowatt hours (KWH)}$$

$$\frac{\text{watts} \times \text{hours}}{1,000} = \text{kilowatt hours}$$

Suppose that an electric iron is marked 8 amps and operated on 110 volts for two hours. What will be the cost of operating this iron for the two hours at 4 cents per kilowatt hour?

$$\frac{8 \text{ amps} \times 110 \text{ volts} \times 2 \text{ hours}}{1,000} =$$
$$1.76 \text{ KWH}$$
$$1.76 \times 4 \text{ cents} = 7.04 \text{ cents or}$$
about 7 cents for the two hours of service

METER READING

The water used in your home flows through pipes from the water mains in the street. To measure the amount of water used in your home, the water flows through a water meter. Water is measured in gallons. Electricity sold to the consumer must likewise be measured. The electric power company installs a meter to measure the amount of current used in your home or school or shop. See Fig. 84. As the

LIGHTING WITH ELECTRICITY

current flows through the meter, the electricity runs a type of motor to turn the hands on the meter dials. The more current used, the faster the current flows and the faster the meter runs.

In Fig. 85 are the dials on a typical electric meter. The four dials are placed to be read as you read any four-figure number. For example, the number 1,490 is read one thousand, four hundred, and ninety. The first dial on the meter gives the *thousands* of KWH's; the second gives the *hundreds* of KWH's; and the last two dials give the *tenths* and *units*.

Only two of the hands on the dials travel clockwise; the other two travel counterclockwise. Notice the direction of the arrows in each dial. The mechanism which controls the dial hands consists of sets of gears run by the "motor." The unit hand goes around ten times, while the tenths hand revolves but once. The tenths hand goes around ten times, while the hundredths hand revolves once, and so on. Always read the number over which the hand has last passed.

Notice the position of the hands in Fig. 86. The thousands hand has passed over 4 and not quite approached 5. The reading is 4,000 for that dial.

The hundreds hand has passed over the figure 7 but has not reached 8. The reading is an additional 700 KWH.

The tenths dial is pointing *at* the two; but has it reached the two yet, or has it gone on by just a little way? Under such circumstances, it is necessary to read the next dial to the right.

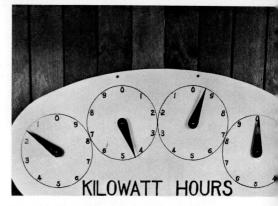

Fig. 85. A practice dial board to learn meter reading — not meant to read correctly as here shown.

Fig. 86. Reading a meter.

Fig. 87. Two monthly readings.

The units dial hand has not quite reached 0 which indicates that the tenths hand must be approaching the two and not on it or past it. Thus, the reading of the third dial is 1 or **10** KWH.

The units hand has passed nine and has not reached the 0; thus it is read **9** without regard to the fraction **or** portion of space between 9 and **0.**

The complete reading of the meter is 4,000, 700, 10 and 9 or, as usually written in arithmetic, 4,719 KWH.

Meters are never reset to the 0000 reading after each reading by the meter man. The number of KWH used in a month or in any given period is equal to the difference in readings between the two periods.

In Fig. 87, take the two readings and find the cost of the electricity used during the month at a flat rate of 6 cents per KWH. The January reading was 9,550, and the February reading is 9,704. The difference is 154 KWH, the amount used between readings. 154 × 6¢ = $9.24, the cost of the electricity for that period.

Compare your findings with payment at home for the last month.

Many power companies use a "step" rate, one in which you pay more for the first "step" than for the second and more for the second "step" than the third and so on. The more electricity you use, the less it costs per KWH. In the following illustration, the rate per KWH is 5 cents for the first 25 KWH used, and 4 cents for the next 25, etc.

Suppose you used 154 KWH during a month.

```
 154
 −25    first 25 KWH at 5¢  = $1.25
 ───
 129
 −25    second 25 KWH at 4¢ =  1.00
 ───
 104
 −50    next 50 KWH at 3¢   =  1.50
 ───
  54
 −50    next 50 KWH at 2½¢  =  1.25
 ───
   4    next  4 KWH at 2¢   =   .08
        ───                    ─────
        Total 154       Total $5.08
```

JOB 19: METER READING

Purpose: To learn to read an electric meter.

Steps:

1. Obtain or make a cardboard meter face showing the dials with movable cardboard hands. See Fig. 85.

2. Have the instructor set the hands to several readings. When you prove that you can make several such readings correctly, he will give permission to proceed to your next job.

3. Read your meter at home; reread the dials a week or so later to see how much has been used.

EXTENSION CORDS

Your next job will be to make an extension cord with an attachment plug on one end and a light socket attached to the other end. The plug cap attached to the one end is connected in the same way as the cap that you installed on your heating-appliance cord.

Fig. 88. Bulbs requiring different size receptacles.

Fig. 89. Types of pendant lamp sockets.

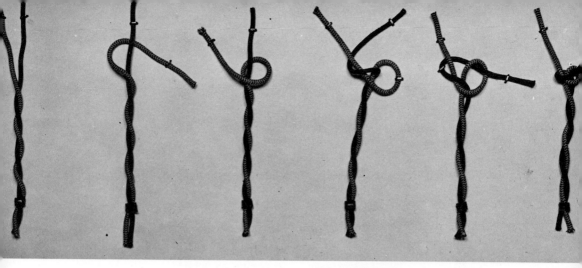

Fig. 90. Steps in tying an Underwriter's knot.

An extension cord assembled from the materials given in this job is recommended for use only in the main part of the home. It should not be used in the basement, for appliances requiring high amperage, out of doors, or in the garage. Extension cords for such places as damp basements or garages must be made of heavily insulated wire and sockets. For such cords, use a wire specified as an all-rubber cord, type S.J.; a rubber-covered socket; and an all-rubber attachment plug cap.

LAMP SOCKETS

There are several types of lamp sockets. Sockets must fit bases of the bulbs. Fig. 88 shows the: (1) miniature, (2) candelabra, (3) intermediate, (4) standard, and (5) mogul sizes of lamps, each requiring a different socket. The sockets may or may not have built-in control switches. The switch is so made that the current is broken rapidly when the bulb is to be turned off. A rapid break of the current prevents arcing. Fig. 89 shows the key type socket, plain socket without a switch; the pushbutton socket; and the pull-chain type socket.

Examine the different types of sockets. See if the amperage the socket will safely carry is shown on each.

JOB 20: EXTENSION CORDS

Purpose: To make extension cords for lighting purposes.

Steps:

1. Obtain a 16-inch piece of lamp cord or silk braided cord (a longer piece may be used if the cord is to be taken home for use); a plug and cap; tape; a pendant standard socket; and necessary tools.

2. Attach the plug cap as you did in a previous job.

3. Practice tying the Underwriter's knot. The knot takes all the strains put upon the cord. Without a knot, the strains are upon the binding-post screws and might pull apart.

For practice use old cord or rope. See Fig. 90.

4. Attach socket as shown in Fig. 91. Remove the cap from the socket by pressing the thumb on the casing where it says "Press." Slip the casing from the plastic or

Fig. 91. A disassembled socket on a cord.

Fig. 92. A common type of feed-through switch.

porcelain base and note where the wires are to be attached. Remove the insulation from the wire ends as before, twisting the small wires tightly together. Screw a small bushing into the cap and fasten lightly with the set screw. Slip the cap onto the cord and tie the knot. Fasten the wires clockwise under the binding screws. Replace the brass casing over the socket and push the cap onto the casing until it snaps into place.

5. Have the lamp cord inspected; then test with a lamp on a 110 volt line.

FEED-THROUGH SWITCHES

Examine two or more types of feed-through switches. The feed-through switch is commonly used on appliance cords as a convenient control. Note the capacity rating of the feed-through switch and always select one which will carry the amperage of the appliance on which it is to be used. See Fig. 92.

JOB 21. THE FEED-THROUGH SWITCH

Purpose: To connect a feed-through switch in an extension cord.

Steps:

1. Obtain a feed-through switch and 10 inches of thread. Use your extension cord from a previous job.

2. Decide the place on the cord where the switch is to be located and mark its full length, as shown in Fig. 93.

3. Wind the cord with 4 inches of thread inside the mark representing the location of the ends of the switch. The thread will keep the outer braid from unraveling.

4. Remove the outer braid between the thread ties. Do not damage the rubber insulation next to the wires.

Fig. 93. Treatment of wires.

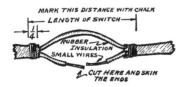

Fig. 94. Parts of the feed-through switch.

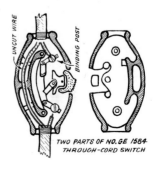

LIGHTING WITH ELECTRICITY

5. Separate the two wires and cut *one* at the center. See Fig. 93.

6. With your knife, skin about ¾ inches of the rubber insulation from each of these ends. Twist the small wires together with your fingers.

7. Remove the assembly screws and take the switch apart. See Fig. 92.

8. Place the uncut wire through the switch as in Fig. 94.

9. Attach each bare end under the binding posts.

10. Replace the cap of the switch and tighten the assembly screws.

11. Test by plugging into the outlet. Then submit your work to the instructor for approval and a mark.

OTHER TYPES OF LIGHTING

There are many other ways of obtaining light from electricity besides that of using the ordinary light bulb.

Arc lighting was once a common method of lighting streets at night. A brief description has been given of obtaining light from an arc produced by the electricity crossing through the air gap between two carbons. Under those circumstances, heat is produced, and light as well. The carbon arc light is still in common use in commercial moving picture machines and searchlights because of the intense light that such an arc gives off.

There are several gases, such as neon gas, which glow when an electric current is present. The gas is enclosed in a glass tube, at each end of which is a metal electrode connection to the space in which the gas is placed. An electric current is made to flow through the gas between the electrodes. Different gases glow in different colors. Both fluorescent and neon lights work on this general principle. The sockets for these lights are quite different from the ordinary light bulb sockets, and the voltage required is also different.

Small tubes, one-half inch in diameter and four inches long, filled with one of the rare gases, such as Xenon, have been introduced. These give off a very intense light and are replacing many arc-lighting installations, to provide an operation free of smoke and fumes.

Since the gases in fluorescent tubes are dangerous, be very careful in handling them. A fluorescent fixture in common use has a starter cartridge which is removed like a bayonet base bulb. These starters are rated in watts. A flickering light may be due to a defective starter. If the tubes and starters are not defective, then it is usually necessary to replace the ballast. Ballasts are numbered, so replace with one having the proper number. The wiring diagram is shown on each ballast.

EXTRA JOB *A:* To make a parallel-circuit Christmas-tree light string. Cut the sockets (leaving at least 6 inches of lead wires to them) from an old series string of Christmas-tree lights. Test each socket with a bulb, using a 15-volt source. Remember that 15-volt light bulbs are used in series-wired cords. For parallel circuits, 110-volt bulbs will have to be used, unless a transformer is included in the circuit and set at the same voltage as the bulbs. Obtain a 10-foot piece of green silk-covered lamp cord, and wire the sockets in parallel to the cord. Connect one socket to the end of the lamp cord and an attachment plug to the other. Make connections for each socket about a foot apart. Solder and tape the connections.

EXTRA JOB *B:* Make an enlarged sectional view of a socket; color the conductors red and the insulators blue. Color those parts black which do not conduct a current but are not insulators of the circuit.

EXTRA JOB *C:* Make a collection of as many types of electric light bulbs as you can find. (Use burned-out bulbs.) Or make a collection of different kinds of electric wire. Mount on a panel and classify each.

EXTRA JOB *D:* Examine and study a fluorescent lamp. Make a drawing of the circuit; prepare a five-minute talk and demonstration.

EXTRA JOB *E:* Visit a local movie theater and get permission to see the arc lighting in the lamp house of the movie projector. Some theaters may not use this type of lighting. Consult your librarian and read about arc lights. Prepare a three-minute talk on arc lighting to be given before the class.

EXTRA JOB *F:* Ask your instructor for a short length of 3-wire extension cord, a grounded plug, and an adapter for an outlet, to facilitate the use of the grounded plug.

Attach three wires on one end of the cord to the three contacts, and show the class how the outside wire on the adapter is attached to the screw on an outlet cover plate. Explain how this ground-wire circuit is connected to the frame of an appliance, such as a portable drill. Bring an appliance, furnished with a three-wire grounded attachment cord, and show how it is used with an adapter in any common outlet.

Why is it safer to use a grounded appliance when working outside, or on a damp concrete floor?

QUESTIONS

1. Will adding more lights to a parallel Christmas-tree string lengthen or shorten the "life" of the bulbs?

2. A dry cell with 30 amps has what wattage? What part of a kilowatt is this? At 5 cents per KWH, what is the electricity in the battery worth if it were possible to buy it at the same rate as house current?

3. If a home has 15 lights of 40 watts each, 10 lights of 60 watts, and 3 lights of 300 watts, how many kilowatts are used per hour if all bulbs are turned on?

4. What would be the gross light bill if the meter showed the use of 107 KWH of electricity, and if the "step-rate" plan described in this chapter were used?

5. What would it cost per month to operate a radio that consumes 60 watts of current per hour if the radio were used four hours per day for thirty days, and the rate was 4 cents per KWH?

6. What does a 110-volt, 100 watt, standard light bulb cost in your community?

7. Why do fluorescent lights require a few seconds to come on after the switch is closed?

8. Why aren't carbon arc lamps used in the home for general lighting?

9. Discuss "indirect lighting"?

10. What are some popular ideas for lighting the "ideal home"?

LEARN TO SPELL

filament	fluorescent
tungsten	carbon
toggle	extension
watt	miniature
conduit	receptacle
kilowatt	meter

Chapter IX

HOUSE WIRING : Electrical Conduits and Switches

Kinds of wire used—joints—soldering—knob-and-tube wiring—switches—conduit wiring—bending with a conduit—pulling wires through conduit.

THE GOVERNMENT built a new dam to furnish power for a valley. Mr. and Mrs. Wilson own a farm in this valley. Their home is old, and there was never a power line near so that they could have the conveniences of electricity. Now that the dam is in and power line is to run along the highway right outside their home, they are making big plans. They are going to build a new house which is to be wired for electricity!

That new home will probably stand there from fifty to seventy-five years if nothing happens to it in that time. And the wires are to run within the walls, through the attic, and between the floors and out to the barn, too. Those wires must last the fifty to seventy-five years, and they must be so installed that they cannot possibly cause a fire that might burn down the home.

Would you want the responsibility of doing work that must last for fifty to seventy-five years? The trained electrician who does house wiring will tackle it because he knows how. Most electricians are licensed, which means that they have proved to society that they know how to do acceptable wiring and have been granted a license permitting them to work as an electrician.

The contents of this chapter should convince you that house wiring requires a great deal of further training, as well as years of experience, to do all the jobs that an electrician is asked to do.

For years homes were wired by a method called "open wiring," meaning that the wires were fastened to insulators which held them away from wood or other combustible materials and not run through conduit. Where the wires were required to pass through lath and plates, or flooring, or across other wires, porcelain tubes or fibre tubes called "loom" were slipped over the wire at those places as added insulation.

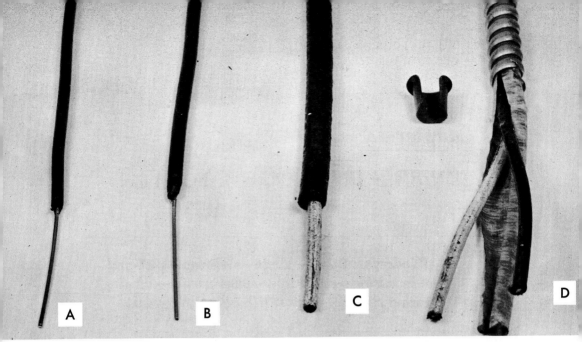

Fig. 95. Wire used in house wiring.

In most communities, "open wiring" is not now permitted. The modern method is to run the wires in steel tubes or flexible metal tubes or fibre sheathing. This method is called conduit wiring. When appliances are turned on do lights dim? If so, wiring may be inadequate for present day loads. Are there two lead wires connecting *your* house with the power line, or are there three? (See Fig. 83.) What voltage does the water heater require? Why? In conduit wiring, the wires are enclosed within the tubing; therefore you must imagine the circuit, or trust to a wiring diagram.

As in bell wiring, it is often necessary to join wires together. The wires used now will be larger than those used in bell wiring, and the joints must be soldered and taped.

WIRE USED IN HOUSE WIRING

Fig. 95, "A", shows various types of wire used in wiring homes. No. 14, solid, rubber-covered, copper wire has a cotton braid over the rubber and is the most common type and size used for ordinary purposes. In Fig. 95, No. 10 R.C. is shown as "B." White, solid, rubber-covered wire projecting from a piece of loom is at "C", and BX cable is at "D".

White wire is used as the connection to the return (or ground) wire. The black wire (referred to as the hot wire) is a feed wire. Alongside and to the left of the BX in Fig. 95 is a fiber bushing that must be installed at the end of the conduit or BX. This will prevent any possibility of a short taking place between the wire and its protective cover.

You may have noticed that the lead wires are usually strung from power poles to homes in sets of three wires. If these wires are lettered "A", "B", and "C", connections to "A" and "C" have a "potential" (voltage) of 220 volts; those to "A" and "B" have 110 volts, and those to "B" and "C",

HOUSE WIRING

110 volts. The various circuits within the house lead to the switch box near the meter, where the source connections are made. A good electrician attempts to estimate the load which may be required of the various circuits within the house and connects these circuits in the switch box so that half the total load will be connected to "A" and "B" and the other half to "B" and "C". You may wish to check this balance in your own home to see whether one side must carry a much greater load than the other. If so, consult an electrician to see if he thinks such a change is necessary. If you have a water heater where 220 volts are required, "A" and "C" would be used.

MAKING JOINTS

The most commonly used splices are the Western Union splice, the tap splice, and the rat-tail splice. Fig. 96 shows how to remove the insulation from wire in the preparation for joining. To remove insulation:

1. Determine the amount of insulation to be removed.

2. Loosen the insulation with a pair of pliers by squeezing the surface all the way around the wire.

3. Hold wire in left hand, Fig. 96.

4. Hold the knife in right hand and at the same time hold the wire between the thumb and the blade of the knife. Take a slicing cut as if sharpening a pencil. *Do not cut into the wire!*

5. Turn the wire, and continue the cutting until the insulation is off.

6. Scrape the wire with the square edge on the back of the knife blade until the wire is bright and clean.

JOB 22. THE SOLDERLESS RAT-TAIL SPLICE

Purpose: To learn the use of solderless connectors.

Steps:

1. Obtain two pieces of No. 14, black, solid, rubber-covered, copper wire, each 8 inches long; side cutters or wire stripper; knife; and a medium-size connector (also called wire nuts).

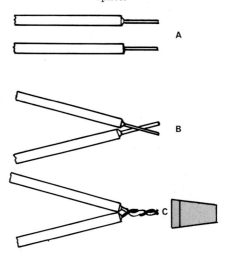

Fig. 97. Making the solderless rat-tail splice.

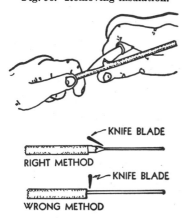

Fig. 96. Removing insulation.

2. Remove 1 inch of insulation from one end of each wire and scrape exposed wire bright. See Fig. 97.

3. Cross the wires as shown and twist tightly together.

4. Screw on a connector until it covers all exposed wire and is firmly in place.

5. Have your instructor approve it.

The stripping tool is described on Workbook pages 45 and 46, Figs. VII-4 and VII-5. Also note the use of this tap, Fig. IX-4, in the Workbook on page 61.

JOB 23. THE TAP SPLICE

Purpose: To learn how to make a tap splice used for branch-circuit leads

Steps:

1. Obtain two pieces of the same wire used in the previous job, each 8 inches long; a knife; and side cutters.

2. Remove about 3 inches of insulation from the end of the tap wire. See Fig. 98.

3. Remove about 1 inch of insulation from the main wire where the branch splice is to be made.

4. Scrape the wires clean and bright.

5. Wrap the branch wire around the main wire, starting as in "B", Fig. 98. Be sure the two bare wires come into contact at point "X", as shown in Fig. 98, "B".

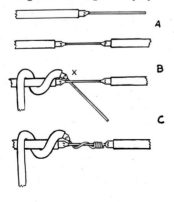

Fig. 98. Making the tap splice.

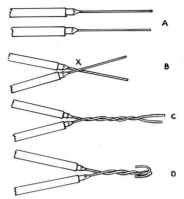

Fig. 99. Making the rat-tail splice.

6. Make two long turns and four short turns, as in "C", Fig. 98.

7. Cut off the surplus wire and press the end down with the pliers.

JOB 24. THE RAT-TAIL JOINT

Purpose: To make a rat-tail joint used in joining two or three wires together in a conduit junction box

Steps:

1. Obtain two pieces of No. 14 solid R.C. wire; side cutters; a knife.

2. Remove about 2 inches of insulation from the end of each wire, as in Fig. 99, "A".

3. Scrape each wire clean and bright.

4. Cross the wires as in "B".

5. With the right hand, twist these ends tightly around each other as in "C".

6. Finish the joint by turning back the ends with pliers, as in "D".

Solderless rat-tails are acceptable if connectors such as those illustrated are used. In Fig. 99, only one half inch of wire is bared, and the turning back of the ends at "D" is omitted. The connectors are screwed onto the splice as far as they go, covering all the bare metal. However, you should learn the soldering process.

Types of solderless connectors.

SOLDERING

Solder is a metal alloy of lead and tin (an alloy is a mixture of two or more metals) which will melt at lower temperatures than the metal conductors will melt, low enough to be melted by a hot flame or hot soldering copper. Solder is bought either in bars or as wire.

Solder will stick to other metals only when a *flux* is applied where the metals and solder are to be joined. Flux is a chemical, used in the form of a liquid, a paste, or a solid. Flux completes the cleaning action of the materials. One variety of wire solder is hollow like a tube, and the hole through the center is filled with a paste flux, such as resin. The type of flux to use depends upon the material to be soldered. Flux is also used when *tinning* (coating the tip with solder) a soldering copper.

Soldering coppers are pointed bars of copper with an attached handle and are heated either by an electric heating unit built into the tool or by the flame from a gas furnace or a blowtorch. The tip of the soldering copper must be coated with solder before drops of fluid solder can be carried and applied to the metal to be soldered. A tinned soldering copper that gets too hot will lose the solder on its surface and will need to be re-tinned.

Tinning Soldering Copper

To tin a soldering copper, follow these directions:

1. Heat the soldering copper; then place the copper in a machinist's vise and file the four sides of the point while the copper is hot until they are clean and bright.

2. Remove soldering copper from the vise and rub the point on a cake of soldering flux while the soldering copper is still hot.

3. Touch the hot copper to the solder, turning the point until all sides have a coat of tin.

4. Wipe the point quickly on a pad of cloth.

5. The copper is now ready for soldering, provided the entire point is completely tinned for more than a half inch.

When possible, the joints to be soldered should be above the copper.

Methods of Heating a Soldering Copper

The gas furnace common to most school shops should be lit with a match or igniter (provided there is no pilot light on the furnace) immediately after turning on the gas. By all means, watch your soldering iron so that it does not overheat. Never go away and leave the copper in the heat. Be careful to keep the iron away from other students.

(The electric soldering iron is heated by connecting it to the cor-

Fig. 100. Heat travels upward.

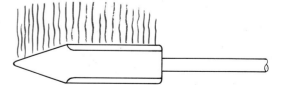

HOUSE WIRING

rect current. Rest the hot end on a metal holder so that it doesn't burn your bench top or start a fire.)

The blowtorch, as a portable heater, may be used to heat soldering coppers. A new torch is usually accompanied by complete instructions. The blowtorch can be dangerous, because gasoline is used for fuel. Before opening the can of gasoline, make sure that no open flames are near; and put the can away before lighting the torch. Wipe off any gasoline that may have spilled over the torch, the bench, or the floor when filling or starting—*before lighting a match!* Also take care of the cloth you use!

Torches smaller than the usual blowtorch are often used to heat the wire and solder directly. No soldering copper is then used. Most of these small torches burn alcohol for fuel. Observe the same precautions in their use as those observed in using gasoline. In many schools, blowtorches and alcohol torches are not allowed.

Also, a small alcohol lamp is sometimes used to heat the wire directly. But an open flame for direct heating has two disadvantages. First, it is dangerous because of the possibility of starting a fire, and second, the flame often burns the rubber and the cord insulation.

A soldering gun is now in common use. It is a fast-heating tool, as it heats at the touch of the trigger. See Fig. 102.

JOB 25. SOLDER AND INSULATE A WIRE JOINT

Purpose: To learn how to solder and tape a joint

Steps:

1. Obtain soldering copper; cake of sal-ammoniac flux; paste flux; wire solder.

2. Use the joints from the previous three jobs for your exercise in soldering.

3. Tin the copper, if necessary, and heat.

4. Apply a small amount of soldering paste to the joint. If resin core solder is used, paste is not. See Fig. 101, "A".

5. Hold the heated copper below (Fig. 100) and touching the joint, for five seconds, at least, before applying solder. Soldering short pieces of wire requires two hands to hold the copper steady, and one hand to hold the solder. This adds up to more hands than you have, so ask another student to hold the wires or clamp in a vise. Apply the solder from above. See Fig. 101, "B". Use just enough to flow smoothly and to seal the joint. Avoid burning the insulation. Have the soldering checked before applying the tape! Important!

6. Rubber tape is applied first, to comply with N.E.C. rule No. 2p3. A piece 2 inches to 3 inches long is sufficient for the average joint. Begin at one end of the joint and wrap spirally to the other, overlapping the *old rubber insulation* as in Fig.

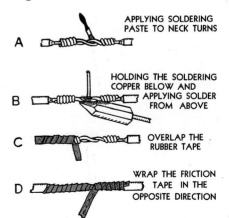

Fig. 101. Soldering and taping a splice.

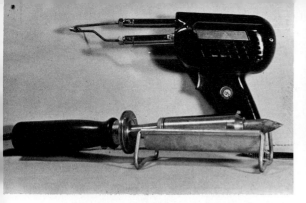

Fig. 102. The electric soldering iron and gun.

Fig. 103. Materials for soldering.

101, "C". Stretch the tape as you wrap, and overlap at least one half the width of the rubber-gum tape. If the tape does not stick, heat it a little until it does.

7. Apply friction tape over the rubber gum as in "D", Fig. 101. Wrap tightly in the opposite direction, and overlap as before, especially at the ends.

8. Turn in to your instructor for inspection and credit.

Fig. 104. To replace wall switch.

COURTESY, WESTINGHOUSE

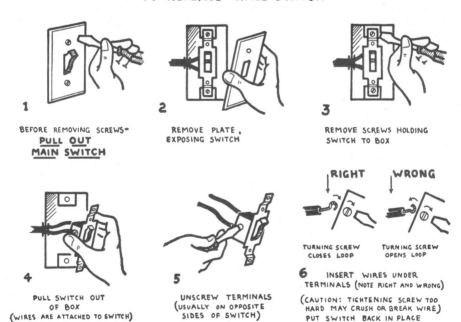

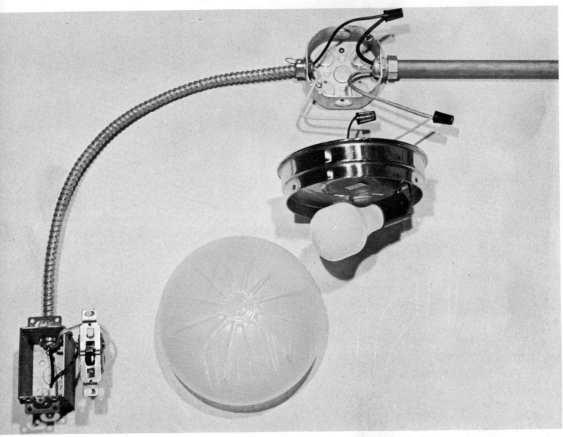

Fig. 105. Adding a wall switch to control ceiling light at end of run.

JOB 26

Purpose: To learn how to connect flexible armored cable (BX) to an outlet box

Steps:

1. Examine carefully the operations shown in Fig. 106.

2. Obtain an 18" piece of ½" BX armored cable, connector, bushing, switch box, and tools.

3. Cut BX about 6" from the end. Hold hack saw at an angle so that you can cut through one section of armor. Be sure not to damage insulation of the wires. Grasp cable with one hand on each side of cut and twist sharply to break armor. Remove this piece.

4. Insert bushing between paper and armor. Notice the bond wire and cut it about 3" long. This bond wire in the cable is to make sure that the cable is a good conductor for grounding.

5. Slip connector (with locknut removed) over wires, bond wire, and armor. Twist bond wire around screw, then tighten screw on armor.

6. Insert the cable (with connector fastened in place) into the knock-out hole in the box and tighten the locknut. (One side of the box has been taken out to show this detail—See Fig. 106 D.)

HOUSE WIRING

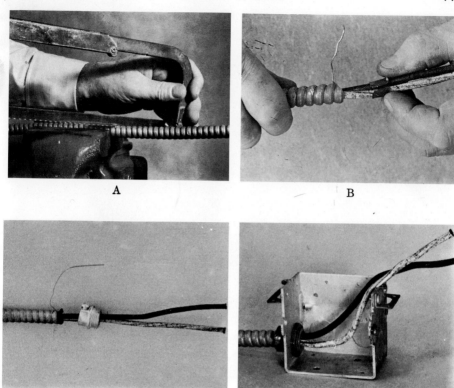

Fig. 106. Steps in using flexible armored cable.

JOB 27

Purpose: To learn about using BX or Greenfield by adding a wall switch to control ceiling light at end of run (See Fig. 105)

Steps:

1. Make a plan of your job and estimate material needed.
2. Obtain necessary tools and materials.
3. Locate switch and ceiling outlet.
4. Make proper knock-outs and fasten boxes in place.
5. Cut cable (See Fig. A, B, C, D, Fig. 106) and connect to boxes. Make sure bond wire is connected under connector screw.
6. Make proper wire connections. Solder and tape all joints or use solderless connectors. Have these inspected by your instructor.
7. Complete assembly of switch, plate, fixture, junction box cover, and bulb. Test.

SWITCHES FOR HOUSE WIRING

To make or break a line circuit in house wiring requires that the connection be made or broken rapidly to prevent arcing within the space where the connections are being made or broken. Switches for turning lights

Fig. 107. Types of switches for the home.

or other appliances on or off are devices to make or break the circuits rapidly. Fig. 107 shows various types of switches used in house wiring. No. 1, is a single-pole toggle switch, one of the most common types. The opposite end of the familiar toggle handle is forced into a coil compression spring. The spring rests on a knob in the cradle of the plastic crank bar shown as "E" in Fig. 108. "E" snaps back and forth with the movement of the cradle handle. The coil spring gives it the snap. In the single-pole-type metal bar, AB connects, or disconnects, from two poles, 1 and 2, like those shown in Fig. 109. Bar CD and contacts 3 and 4 are required only in three and four way switches described later in the text.

No. 2, Fig. 107, is the silent mercury switch; No. 3 is a delayed-action switch, with the current staying on for a few seconds after the switch is turned off. No. 4 is a dimmer switch; No. 5 is the touch-plate type, and No. 6 is the pilot-light and switch, with the red warning light always showing when the outlet controlled by the switch is in.

When it is desirable to control a load such as a stairway light from two or three locations, types 1, 2, and 5 are available for such circuits.

Three-way Switch

In two previous jobs, you wired a light, or a bell, to be controlled by either one of two switches. Well, here it comes again! This time, the switch to be used is a three-way toggle switch. Fig. 108 shows the cradle with two sets of connecting bars, AB and CD. Fig. 109 shows the three-way switch used. Now turn to Fig. 111, showing the simplified, old-fashioned, revolving snap switch with four contacts, the connecting bar revolving to connect opposite poles. Note the dotted line between two poles, representing a jumper wire installed to make a 4-way into a 3-way.

HOUSE WIRING

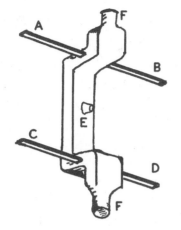

Fig. 108. Cradle of the toggle switch.

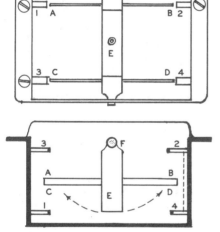

Fig. 109. Top view looking down into the three-way toggle switch, and the end view of same with front plate removed.

Trace the circuit in Fig. 111, page 102, to see how either switch turns the light off or on. The toggle switch, Fig. 109, is made as a three-way, with the contact No. 4 shunted to contact No. 2. Keep in mind that contact post 3 is at the top of the box, and No. 4 is at the bottom; both are on the same side of the box. Likewise,

No. 2 is at the top, and No. 1 is at the bottom and on the other side of the box. "A" and "B" connect and disconnect with No. 1 at the bottom, No. 2 at the top, and CD with No. 3 at the top, and No. 4 at the bottom. In one switch, Post 2 is connected to the line, and in the second switch, it is connected to the lead to the light. The two control wires shown on Fig. 111 are connected to respective posts, the No. 1's together, and the No. 3's together.

Fig. 112, page 103, shows the four-way switch, a third switch to make it possible for a light to be controlled from three locations. Here a four-way toggle is used for the center switch. In the four-way toggle, the jumper wire from 4 to 2 is omitted, and a contact post for No. 4 is included for it. The switch is used as a four-way switch.

Mercury Switch

Another type of switch, shown in Fig. 107, No. 2, that is rarely found in the usual light circuit but is used for many other purposes is the mercury-type switch. It has the advantage over the snap or toggle switch in that it is noiseless. See Fig. 113, lower.

Mercury is a metal which is in liquid form at ordinary temperatures. At 40° below zero it becomes a solid, the state in which we recognize most other metals. Mercury is a good conductor of electricity, so one of its uses is in the mercury switch.

The mercury switch is simply a glass bottle which is partly filled with mercury. At one end of the bottle, there are two wires molded into the

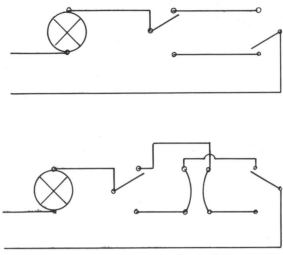

Fig. 110. Wiring schematics for the following two figures.

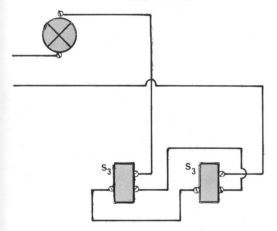

Fig. 111. Lights controlled by either one of two switches.

glass. The ends of the wire contacts are connected outside of the glass container to the control and the feed wires. The other two ends of the contacts are inside the bottle, so that when the bottle is tipped one way the mercury runs to the end of the bottle and around the contacts to complete the circuit. When the bottle is tipped the other way, the mercury runs away from the contact, breaking the circuit. See Fig. 113.

The switch mechanism is a simple device to tip the bottle one way or the other.

Thermostatic Switch

The thermostatic switch is a device to open and close a circuit by automatic action caused by changes in temperature. The thermostat is very

HOUSE WIRING

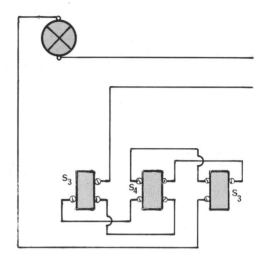

Fig. 112. Lights controlled by any one of three switches.

commonly used to turn the heating furnace on or off as the temperature of the air in the room where the thermostat is placed changes.

Metals expand when heated and shrink when cooled. A steel flag pole is taller in summer than in winter! A copper strip that measures exactly 12 inches long at 40° below zero would be more than 12 inches long at a temperature of 100° above zero. Now, some metals expand much more than others. Suppose you have two thin ½-inch strips of metal, each 12 inches long. One strip is made of a metal that expands a great deal when heated, and the other strip is made of a metal that expands but very little when heated.

Now suppose that you "fuse" or cement the two strips together as shown in Fig. 113, top, so that you have one "bimetal" strip, twice as thick as before but remaining ½ inch wide and 12 inches long. If you were to place the strip on a hot stove, the strip would "warp" to the shape of a rocker on a rocking chair. One flat side of the bimetal strip would be longer than its other flat side, because half the strip has expanded more than the other half. Fig. 114 shows the bimetal strip anchored at a coiled end; the other end is left free to move from left to right and to touch either contact. The neutral position may be reduced almost to zero by tightening the two adjusting screws, if extreme sensitivity to heat is desired. Trace the circuit to see how the thermostat automatically controls either one of the two lights.

Entrance Switch

The main wires that enter the home of a consumer (Fig. 83) are connected to the main or entrance switch. The switch is built into a steel protective box. The main fuses are also included in the box. See Fig. 115. All circuits must be connected to an electric meter which measures the electricity used in the home.

There is one type of entrance

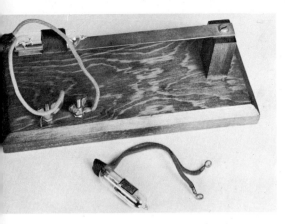

Fig. 113. The mercury switch. The top one is mounted on a bimetal strip.

switch which has no fuses. The circuit is broken automatically within the mechanism of the switch when an overload occurs. Like a toggle switch, the contact arms are held in an "off" position by a spring. When in the "on" position, the spring pulls at the

Fig. 114. Bimetal coiled strip used as a switch. When cold, the left light burns; when warm, the left light goes out and the right light comes on.

contact arms but the contacts are held in the "on" position by the end of a strip of bimetal (Fig. 114). This bimetal "warps" out of position when heated. Since the bimetal strip is part of the circuit, an overload heats it, warps it out of normal position, and thus releases the switch contact arms. This permits the spring to pull the arms back to the "off" position.

There are many, many other types of switches, but space does not permit the description of them here.

Low Voltage Control of Lights

Wall switches for controlling lights in ceiling or on walls have had an interesting development. First, there was the snap-switch; then the push-button-switch; after that, the toggle-switch; this was followed by the silent-mercury toggle switch; and now, the remote control relay type.

This latter type, the relay switch, has a few advantages over the others, which we shall see later. The 110-volt circuit is completed or broken

Wall thermostat, with night-time lower heat setting.

COURTESY, MINNEAPOLIS HONEYWELL

HOUSE WIRING

COURTESY, SQUARE D CO.
Fig. 115. Entrance switch in box.

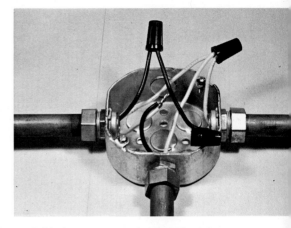

Solderless connectors in a junction box.

within an outlet-box by the relay switch. A small 24-volt transformer furnishes the current to the magnets in the relay. These magnets trip a mechanism to make or break the 110-volt circuit. The low-voltage control wires for the magnets need no more protection than the light kind of wire used in wiring door bells. Thus, no expensive conduit and switch outlets are required for the control circuits. The switches are often mounted right on the wall surface.

See the two figures showing the relay in an outlet box. There are several types of such relays, but only one is described here. The schematic drawing shows the circuit. To turn on the light "L", press button "A". The coil "G" in the relay pulls the armature toggle "E" to close the 110-volt circuit. The toggle remains in this position by spring action until the opposite coil is connected by the other button. Then, to turn off the light, press button "B". Coil "F" now tips the toggle to the other side, and the 110-volt current is now broken.

CONDUIT WIRING

There are four principal types of conduit used in house wiring; they are the rigid, thin-wall, armored cable, and nonmetallic sheathed cable.

A typical transformer. Used here to step 2300 volts down to 220 volts for use in neighboring homes.

COURTESY, WESTINGHOUSE

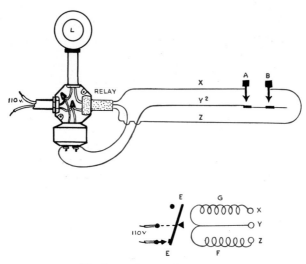

The low voltage relay.

Rigid Conduit

Rigid conduit (Fig. 121) is a special grade of steel pipe installed in partitions or under floors. The conduit is often held in place by pipe straps nailed to the studs or joints. In large steel and concrete buildings, the conduit is embedded in the concrete when the concrete is poured. Wires are pulled through the conduit during the finishing of the building. To lengthen rigid conduit, it must be threaded at the ends and connected by couplings as in Fig. 122. At outlets and switch points, the conduit is connected to metal boxes with special fittings. Wire is "pulled" in after the conduit system has been completed. The fixtures are then installed.

Thin-wall is a type of rigid conduit, shown in Fig. 117, which, as the name implies, is thin steel tubing. Thin-wall differs from rigid conduit in that it does not have to be threaded at the joints. Special clamp-type fittings are used for joining the ends or for joining thin-wall conduit to outlet boxes. See Fig. 117. Thin-wall can be bent more easily than rigid conduit. Bending is done by a hickey or a bender. See Fig. 121.

Armored Cable or Flexible Conduit

Armored cable, sometimes called "Greenfield," is a flexible tubing. Boxes and outlets are installed first and then connected with the flexible armored cable. Wires are "pulled" in after completion of the installation. A box fitting for armored cable is shown in Fig. 118. One type of flexible cable, called BX, is manufactured with the wire inside the tubing. See Fig. 95D. BX is usually restricted in use to short runs and for rewiring of old homes. A nonmetallic covered cable is being widely accepted, especially for wiring farm buildings, and is sold

HOUSE WIRING

COURTESY, LASCO INDUSTRIES, CALIF.

Fig. 116. Plastic conduit: Tubing, boxes, connectors, adaptors, are all plastic. Joints are cemented. Tubing is cut with a hack saw. For over ground or under ground work.

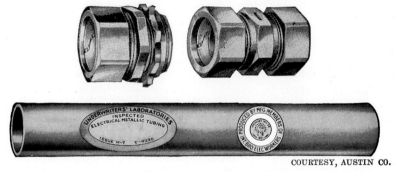

COURTESY, AUSTIN CO.

Fig. 117. Thin-wall conduit and fittings.

under such trade names as Romex or Lomes. This cable consists of two or three strands of heavily insulated wire, wound with heavy paper and covered with fabric.

Romex is installed very much the same as BX. It is fastened to boxes and other fixtures by clamps. No special tools are required. All pieces must be continuous from box to box. Splices must be made within boxes. The cable is anchored by straps, never nails or staples, at distances not to exceed 54 inches, and within a foot of each box.

Conduit is held to a concrete stone or tile wall by metal straps. Metal strips are fastened to such material by toggle bolts or screws turned into metal or wood plugs. The holes for the plug, the stone, concrete, or tile are "drilled" with a star drill. Ask your instructor to show you one.

Outlet Boxes

Conduit always leads to some sort of metal outlet box, such as a rectangular switch box or an octagonal junction box. See Fig. 119. Circles are pressed into the steel sides and bottom of such boxes so that a hole for the end of a piece of conduit may be made by knocking out one of the circles with a hammer.

Now that you are acquainted with the fittings for conduit work, you are to do some exercises using such materials. To understand where the boxes and tubing are located in an actual job, observe such wiring the next time you are in the basement of a building where conduit has been used. You may not see the conduit within the walls, but you can better imagine them from your experiences in this course.

Conduit wiring of a home usually

requires flexible conduit within the walls or between the floors and rigid conduit upon such surfaces as concrete walls or unplastered basement ceilings.

Your first conduit job will be to use rigid conduit between surface outlets and a fuse box, to represent wiring done upon walls.

Cutting Conduit

Rigid black and Greenfield conduit may be cut to proper lengths by clamping in a pipe vise and cutting with a hack saw, as in Fig. 120. After rigid conduit is cut, a burr or sharp edge is left around the inside rim of the tube which must be removed with a reamer, Fig. 122. If no reamer is available, try using a small round file. If the burr is left on, the insulation on the wire may be cut when the wire is being pulled through the conduit.

Bending Conduit

The conduit bender shown in Fig. 121 is a tool used in bending conduit. There are several types. Learn to use the one in your shop correctly.

Threading the Ends of Rigid Conduit

Threads are cut on the ends of pipes by a tool called a die which fits into

Fig. 118

COURTESY, AUSTIN CO.
Fig. 119

Fig. 118. Fitting for armored or flexible cable.
Fig. 119. An octagonal box.

a die holder. See Fig. 122. Your instructor will show you how to select the proper die for the size of pipe you are to thread. Use a cutting oil on the pipe end when cutting threads, so that die is not ruined. Oil reduces the friction and prevents excessive heat from changing the temper of the steel and breaking the cutting threads.

Pulling Wires Through Conduit

Wires are pulled through conduit after the conduit is attached to the various metal boxes. See Fig. 123. This kind of wire is called "fishing wire"; it is made of tempered steel and is rectangular in shape. A small hook is bent on the end of the fishing wire so that it will slide readily past small obstructions and around bends. Oftentimes, just a length of heavy galvanized wire is used as a fish wire. After the fish wire is pulled through, attach the conductor to be pulled in by hooking or looping the wire onto the fish wire; wrap with fine copper wire and cover over with a layer of tape. Pull wire from outlet box to outlet box, leaving 6 inches of extra wire at each box.

Fig. 120. Cutting rigid conduit.

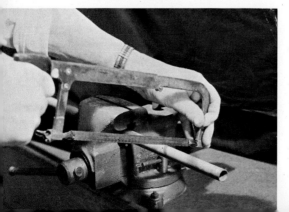

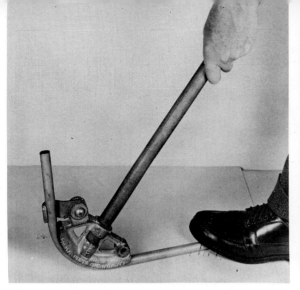

Fig. 121. Bending rigid conduit.

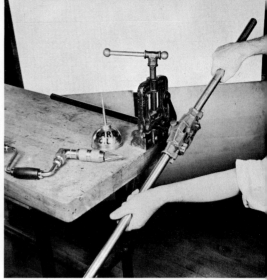

Fig. 122. Threading rigid conduit.

JOB 28. MAKING TWO SEPARATE RUNS OF CONDUIT

Purpose: To learn how to use various types of conduit (See Fig. 124).

Steps:

1. Obtain a fuse box, two switch boxes, two octagonal boxes; one piece of black rigid conduit; one piece of conduit threaded on both ends, and four connectors; one piece thin-wall bent as per diagram, with two connectors; one piece Greenfield with connectors; one octagonal box, cover plate; one octagonal receptacle; one S1 and plate and one switch box, double outlet and plate; and the required amount of No. 14 R.C. wire.

2. Install the empty boxes and the conduit. (Your instructor may permit you to cut, form and do necessary threading of all the pieces. You should have the experience, at least, of cutting off one length of flexible conduit for one connection.)

3. Cut the necessary lengths of wires and make the connections. In box "X", make a rat-tail splice. Do not put on the cover plates until the work is inspected by the teacher.

4. Install plates and test.

EXTRA JOB *A:* Using approved armored cable, boxes, etc., wire a light to be controlled by either one of two switches. Follow diagram, Fig. 111.

EXTRA JOB *B:* Suppose you were requested to modernize the knob and cleat wiring illustrated in Fig. 110. Using sheathed cable and proper materials, wire the circuit shown.

Fig. 123. Pulling wire through conduit. Note switch box.

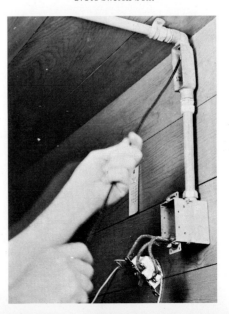

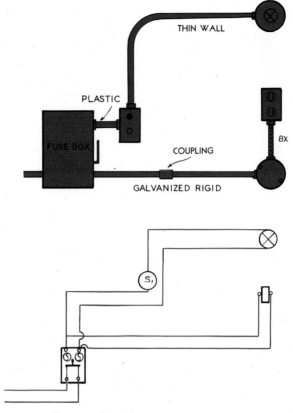

Fig. 124. Conduit wiring and circuit.

EXTRA JOB *C:* Make the thermostatic switch and connect as shown in Fig. 114.

EXTRA JOB *D:* Using thin-wall conduit, or nonmetallic sheathed cable and fittings, wire a light to be controlled by either one of two switches.

EXTRA JOB *E:* Obtain a catalog of house-wiring materials. Plan a light circuit from the house to the garage, with a switch in the house and one in the garage, either of which will control an outside light over the garage entrance. From the catalog, write out an order for all the materials needed. Figure the cost of materials. See Fig. 125.

SYMBOLS

$S1$ = Single-Pole Switch
$S3$ = Three-Way Switch

Ceiling Outlet   Wall-Lamp Receptacle

Convenient Outlet   Junction Box

HOUSE WIRING

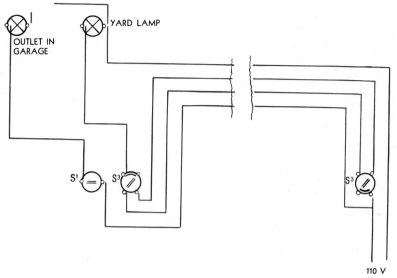

Fig. 125. Wiring diagram of house-to-garage circuit.

EXTRA JOB *F:* Using exposed sheathed cable, wire in outlet box, controlled by a switch from a switch box and originating from a junction box. Strip the cable at least 6 inches from each box. (All exposed cable is strapped every three feet.)

EXTRA JOB *G:* Make a safe test panel for troubleshooting.

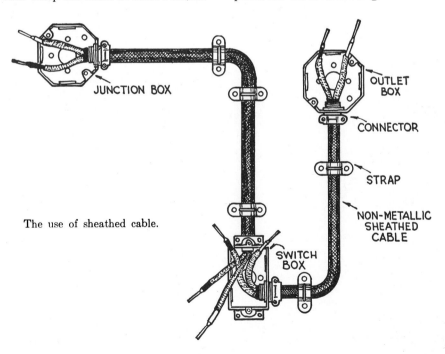

The use of sheathed cable.

INDUSTRIAL-ARTS ELECTRICITY

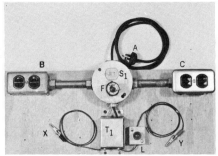

Portable power supply panel for use in the home.

The picture above is a plug-in test panel. Select materials needed and assemble as in the two pictures. When completed, the panel must pass inspection by your instructor before use. Toggle switch S_1 is in series with the attachment cord A. The fuse F (5 amp.) and its outlet are also in series with the switch. Current entering outlets B and C and the transformer must first go through the fuse and switch. Test probe clips X and Y, from the 15V output of the transformer, are in series with the 15V light at L. These probes are used in testing for open circuits.

The outlets B and C are provided for tools needed in troubleshooting. An electric soldering iron and a light are two such tools. The low amperage fuse provides a measure of precaution.

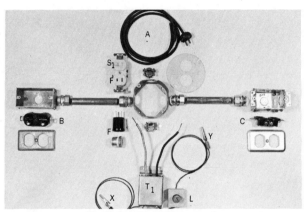

Parts for the power panel.

QUESTIONS

1. Why is "open wiring" not permitted in most communities?
2. Which type of splice do you think is used the least in conduit wiring of homes?
3. Why are electricians who wire houses licensed?
4. In wiring an old house, how do electricians get the flexible conduit down between the walls? Under the floors?
5. How does a thermostat control the heat in a room?
6. What are the insulation bushings for BX Cable? (Consult a catalog.)
7. What are the knock-out plugs in an outlet box?
8. What is a reamer, and where is it used in conduit wiring?
9. What is the difference between a junction box and a switch box?
10. Why should a fuse block be placed in a steel box?

LEARN TO SPELL

conduit	insulation	thermostatic
toggle	mercury	octagonal box
solder	installation	connection
bimetal	nonmetallic	armored cable

Explaining the circuits in a television receiver.

Tracing a circuit in a receiver.

TV troubleshooting.

COURTESY MINNEAPOLIS
VOCATIONAL HIGH SCHOOL

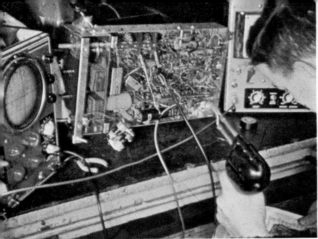

Soldering a defective circuit.

COURTESY MINNEAPOLIS VOCATIONAL HIGH SCHOOL

Checking the signals with an oscilloscope.

COURTESY MINNEAPOLIS VOCATIONAL HIGH SCHOOL

TV repairing.

COURTESY MINNEAPOLIS VOCATIONAL HIGH SCHOOL

Chapter X

COMMUNICATION by Means of Electrical Transmission

Blinker lights—the telegraph, and telegraph codes—the telephone—radio—use of electricity for other means of communication.

SUPPOSE your instructor comes before the class with soot on his nose; he does not know it, but you do. So does your friend at the end of the row. You look at your friend and smile. Your friend winks back. You and your friend are *communicating*.

People transfer their thoughts, ideas, opinions, and information to others in many ways.

Communication is accomplished through one or several senses. An Indian *saw* smoke signals; the savage *heard* a drum beat, the blind man *felt* raised bumps on paper; the rescue plane pilot *saw* greenish color in the water; the dog *heard* a siren; the broker *looked* at a tape; the operator *listened* to a sounder; a surveyor on one hill top *read* the code of mirror flashes made by another surveyor upon another hill two miles away, etc. All these illustrate the reception of messages by the brain. To understand those messages, one must know what the message means to the person sending the information.

Electricity has become one of the most widely used means of communication between distances too great for the ordinary voice to carry or for the eyes to see.

A boy in St. Louis is at the telephone talking "long distance" to his father in Los Angeles. The boy's mother is with her boy in the same room, twelve feet away. The father hears his son's greeting before the mother does! Sound travels 1,136 *feet* per second; electricity travels 186,000 *miles* per second. Electricity carries the message from St. Louis to Los Angeles more rapidly than sound carries the message a distance of twelve feet.

In this chapter many of the methods of communication using electricity are described. We communicate by light from electricity and by sound converted by electricity.

Language, written or spoken, is our common means of communicating with one another. To talk to one another, we must know the same language. The spoken language consists of word-sounds. We also use sounds other than words to communicate, such as dots and dashes. And we use symbols or signs interpreted by the eyes to communicate, such as printed words, smoke puffs, photographs, drawings, or light flashes.

BLINKER LIGHTS

Though Tom's house was a block away, Bill could see the bedroom window in Tom's house. The two boys learned the Morse code. Each had a flashlight, and at night they sent messages in code back and forth to each other. At first they checked their messages over the telephone, but later they became so proficient that they no longer needed the telephone. They were learning code so that at summer camp where no telephone service was available they could communicate at night for quite a distance.

All large fighting ships carry blinker lights to signal by code, either by day or by night, to other near-by ships or stations when radio transmission must be silenced. Land armies have men with each unit who are trained in the use of signaling lights. Large airports have traffic towers where light "guns" are used to signal by colors to pilots who wish to land on the field. The light gun is merely a strong spotlight.

Blinker lights vary in size and usually have powerful lamps, with lenses to strengthen the beams of light. A rapid, smooth-action shutter is operated to cut the light beam into dots and dashes.

THE TELEGRAPH

How Telegraph Operates

As you know, Samuel B. Morse invented the telegraph in 1844 and sent the first long-distance message by electricity. His telegraph was built very much like the one described in this chapter, an electromagnet with a rocker arm which "clicked" each time a current was permitted to flow through the circuit, then "clicked" again when the circuit was opened. A spring pulled the arm up and away from the magnet core when the circuit was broken. The device to make and break the circuit is called a key and is operated upon the principle of the pushbutton switch.

Study the circuit in Fig. 126 to see how the current travels and why the receiving switch must be closed for the sender to send messages. You will notice that one wire is "grounded." The earth is a conductor because of the water and minerals therein and replaces one wire. Using a ground saves the cost of installing the extra wire.

When the key is pressed down (which is the same as closing a switch), the current flows through the circuit. The electromagnets in the sounder attract the metal bar to make the "click." Dots are made by short contact with the key, and dashes are made by holding the key closed for a longer period.

One can understand that to send a current over a long distance the

COMMUNICATION

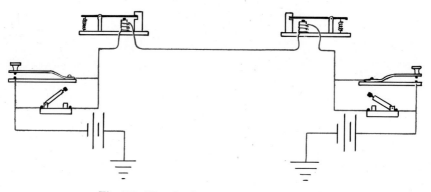

Fig. 126. The simple two-way telegraph circuit.

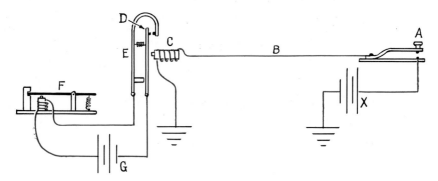

Fig. 127. Relay circuit.

strength of the current must be great if the sounder is to "click" sufficiently loud to be heard. The telegraph *relay* was invented so that messages could be sent farther with the same amount of current. The arm on the telegraph which makes the sounds is fairly heavy and requires more current than a very light arm. The relay is a "sounder" with a very light arm, and upon this arm is a contact which completes or breaks *another* circuit. The relay "takes" the message and passes it on to the next sounder. Fig. 127 shows a relay circuit used in one end of a simple telegraph system. The sender, perhaps a mile or more away from the receiving end, presses the key "A", completing the circuit through the magnet "C" in the relay, the current returning to the battery "X" through the ground. When the relay arm "D" is attracted to magnet "C", the sounder circuit is completed by the contact at the end of the relay arm "D". The battery "G" furnishes

Fig. 128. A telegraph key, a relay, and a sounder.

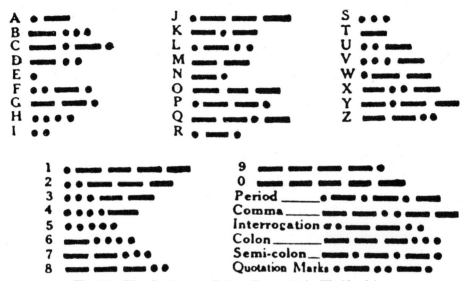

Fig. 129. The Continental Code. (See p. 72 in Workbook.)

plenty of force to make the sounder "F" heard easily. Spring "E" pulls arm "D" back to break the sounder circuit when the sender releases pressure on the key "A".

The Continental and Morse Codes

Morse Code is used primarily in railroad communications. Continental Code, used in radio communications, is received as buzzes from an oscillator. (See pp. 69 and 70 in the Workbook.) The Continental Code is learned as dits and dahs, which replace the Morse Code dots and dashes. (See Fig. 129.)

Bad weather often interferes with radio voice reception. Signals from radio oscillators, however, are seldom blanked out by weather.

JOB 29. THE TELEGRAPH KEY

Purpose: To learn how to make a simple telegraph key. See Fig. 130.

Steps:

1. Obtain the base, ⅜" x 2½" x 6"; a strip of IC tin (or 20-gauge brass) ¾" x 5"; two contact posts; a knob or half a spool; necessary tools and wire.

2. Following the drawing, cut the channels in the base with a small veining tool or gouge; drill the holes in the base and the metal strip.

3. Force a 1-inch brad and the unin-

COMMUNICATION

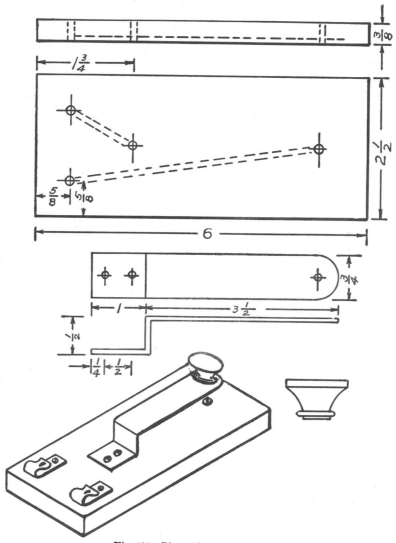

Fig. 130. Plans of a telegraph key.

sulated end of one wire up through the bottom. Wind wire around brad within the hollow part. Apply flux to the wire and the nail.

4. Drip solder with a soldering copper into the opening and, when the solder is cooled, cut off the brad to allow 3/8 inch to project above the top of the base.

5. Install the knob, the posts, and the strip; complete the wiring.

6. Give the base a shellac finish after cleaning all parts and surface.

JOB 30. THE TELEGRAPH SOUNDER

Purpose: To learn how to make a telegraph sounder. See Fig. 131.

Steps:

1. Obtain the mounted electromagnet you made in Job 5.

2. Obtain a piece of ⅛" x ¾" strip iron 2¾ inches long; cut the ³⁄₁₆-inch notch halfway through with the lower edge of notch 1⅝ inches from bend mark with a hacksaw and a file; bend the opposite end ¾-inch in; file the top corners smooth; and drill a ⁹⁄₆₄-inch hole in the bent end. Use a ¾-inch No. 6 screw to fasten.

3. Cut two pieces of ¹⁄₁₆-inch strap iron ½" x 2"; drill two holes ³⁄₁₆-inch in each, center at the end and bend ½-inch of one end at right angles; fasten with ½-inch No. 4 RHB screw. Round the top ends.

4. Cut one piece of ¹⁄₁₆-inch strap iron ½" x 5⅞" for rocker arms; notch one end as shown; drill a ³⁄₁₆-inch hole 2⅛ inches from the notched end; clamp in vise and twist a quarter turn at the center with a monkey wrench.

5. Install rocker arm with a ⅛" x ⅜" stove bolt.

6. Use a cup hook to hold a rubber band by one end.

7. Wire the circuit and then give a shellac finish.

Because wires and telegraph poles were expensive to install over great distances, a method was developed of sending many messages by different parties over the same wire without their interfering with each other.

The telegraph has been in use for a century now, and it is still an important means of sending messages. Lines extend all over the globe, and even across oceans, where cables are laid upon the ocean floor.

In the wireless telegraph, "oscillators" (os'-sil-lators) send out over the air a radio wave which, when picked up, may be heard over radio phones as a tone of a given pitch. These tones are broken into dots and dashes by the sender to be interpreted by the listener. You, no doubt, have heard these on the "short wave" of the home radio. The two letters y-i

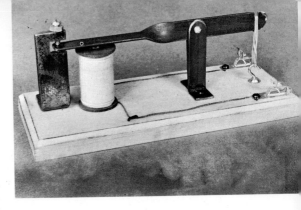

Fig. 131. A simple telegraph sounder.

sound like dah dit dah dah (short pause) dit dit.

THE TELEPHONE

Sound travels in waves similar to the waves that travel on the surface of a pond on a quiet day when one throws a stone into the pond. From the center, where the stone disappears into the pond, the waves move in circles away from the center; these circles become larger and larger in diameter until they finally "disappear." A church bell rings, sending out sound waves in all directions. A person standing within hearing distance "hears" the bell. That is, the waves hit his ear drums; and the nerves from the ear to the brain carry the message, which is interpreted by the brain as a "bell ringing." Different-sized bells, or different kinds of noises, send different waves that our brain learns to tell apart and to "understand." The sound vibrations or waves vary in number from twenty per second up to fifteen or twenty thousand per second!

If someone is beating a drum or playing loudly on a piano and you are sitting near with an empty cardboard hat box on your lap, you may feel

COMMUNICATION

the vibration of the hat-box top with your finger tips. The sound is being "received" by the cardboard box!

The Transmitter

The telephone consists of the transmitter and the receiver. The transmitter receives the sound waves and sends the *effect* of the waves to the receiver by electricity. But the sound waves are not sent over the wires.

You have learned that carbon is a conductor of electricity. In the transmitter, there is a small round box in which carbon granules (small pieces of carbon) are placed. One side of the box is fastened to a diaphragm, or metal disk. When sound waves, as shown in Fig. 132, hit the disk "A", the disk vibrates. The movement of the disk moves the cover "X" of the box "B" and presses on the carbon particles. An electric current (D.C.) flows to the box cover "X", through the carbon, and out of the box at "C", to vary the **pressure of current** passing through.

A noise such as a shot from a popgun pushes the carbon together to permit a good contact; but immediately the box cover returns to its normal position, the carbon particles loosen, and the current flowing through is lessened.

The sound waves, then, are changed to an electric current which changes in strength according to the types of the sound waves hitting the disk.

The Receiver

The job of the receiver is to change this varying current flow back to sound waves.

Fig. 133 shows the telephone receiver. "A" is the disk, or diaphragm, which is to be made to vibrate and thus make sound waves so that the ear may hear.

"B" is an electromagnet connected to the circuit from the transmitter. The force of the current traveling through the electromagnet "B" varies with the different sound waves hitting the transmitter diaphragm. The change in strength of current to the magnet changes the strength of the magnetism in the magnet and attracts the receiver diaphragm with varied "pulls." Thus the diaphragm "A" in the receiver is made to vibrate exactly as the sound waves make the diaphragm in the transmitter vibrate! (The core of the magnet is a permanent magnet to hold the diaphragm in place.)

Complete Circuit

Fig. 134 shows the complete telephone circuit. You will notice that an induction coil is included in the circuit to "step up" the voltage in the line, because actually the voltage changes caused by the sound waves

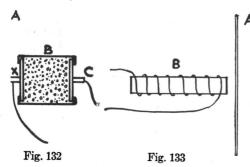

Fig. 132. The transmitter mechanism.
Fig. 133. The receiver mechanism.

Fig. 132   Fig. 133

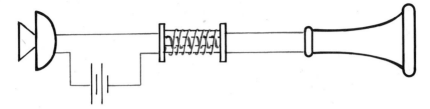

Fig. 134. The simple one-way telephone circuit.

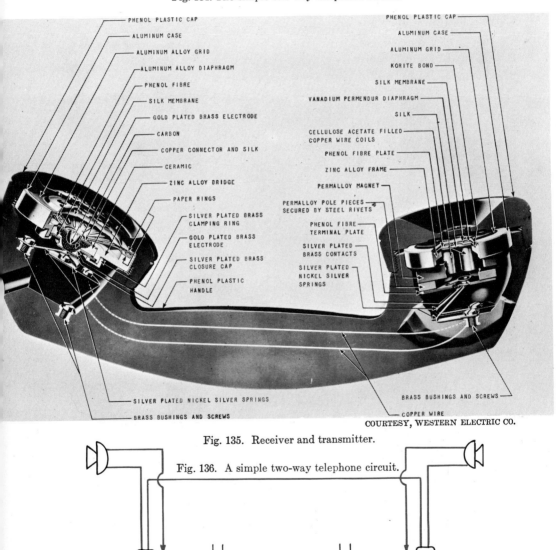

Fig. 135. Receiver and transmitter.

Fig. 136. A simple two-way telephone circuit.

on the transmitter diaphragm are very small.

Study the various parts in the phantom view of a modern telephone set in Fig. 135. Trace the current in the simple two-way telephone circuit in Fig. 136.

JOB 31. THE SIMPLE TELEPHONE CIRCUIT

Purpose: To learn about a simple telephone circuit. See Fig. 134.

Steps:

1. Obtain a telephone transmitter; a receiver; an induction coil; a battery switch; two dry cells; 80 feet of annunciator wire; another student to help.
2. Fasten the transmitter and coil to a panel and install it near one end of the shop.
3. Install the receiver 30 or 40 feet away from the transmitter panel and run the wires between the two. (Since these are temporarily installed, be sure to keep the wires out of the way of the rest of the class.) Then test the circuit.
4. Have circuit checked by instructor; disassemble and return apparatus.

The Automatic Telephone

One may easily see the complications arising when connecting telephone circuits in which more than two telephones are used. Suppose "A" and "B" and "C" each have one telephone. The circuits must be so wired that "A" can talk to "B" or "C", or "B" can talk to "C". Now suppose a fourth party, "D", desires a telephone. "A" wants to talk to "B" or "C" or "D"; "B" wants to call "C" or "D"; or "C" wants to talk to "D". A switchboard becomes necessary as the number of subscribers increases, and someone is employed to "run" the switchboard, to make the connections desired by the various parties. Today, if one has a telephone, he may talk to almost anyone, anywhere.

The automatic telephone switch is a complicated machine. It is used to make the connections automatically by "dialing" the number with a dialing device on each telephone instrument. Fig. 137 shows one instrument which makes a hundred different connections. In a city where two letters and four numbers are dialed for each call, several of these instruments are connected automatically to complete the call—four called selectors, one connector, and one line finder. Hundreds of these instruments may be in one telephone building.

In Fig. 137, you will see the two banks or sets of points. The contacts

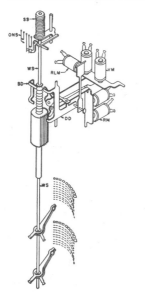

Fig. 137. The automatic switch mechanism.

VERTICAL, ROTARY AND RELEASE PRINCIPLES
COURTESY, AUTOMATIC ELECTRIC CO., CHICAGO

are arranged in rows of ten and are on ten levels, each one of which is the connection to someone's telephone. The wiper shaft (WS) moves up and down as well as around. Notice the lifting magnets (VM) which push the shaft up from one to ten notches. You dial the number "5". Each click of the dial gives an impulse to the elevating magnet which pushes the shaft up to the fifth row in each of the two banks. Notice the two contact arms, which also rise to the fifth level. You now dial "7". The rotating magnets (RM) push the shaft around to the seventh notch, and the contact arms revolve or turn to the seventh contact in the fifth row from the bottom of each bank. The party whose telephone number is 57 is now connected, and an automatic device rings his telephone. The spring (SS) at the top of the shaft swings the shaft back to a "neutral" position, which then allows the shaft to drop to the starting position when the parties are through talking.

It is predicted that a new electronic switch machine will eventually replace all the automatic switches described above. The dialing may be replaced by a set furnished with ten push buttons.

There are many fields of work at which men earn their living which are concerned with the use of electricity. In fact, one could spend a lifetime studying and working in one field alone, such as telephone communication, and never learn all that is to be known about the single field and its many new developments.

RADIO

A few basic facts about radio:

During the hot summer, you probably have seen heat waves "radiating" from a paved highway. Heat is also a form of *energy* just as is electricity. A hot object such as a stove "radiates" heat. You touch a hot stove, and the heat is conducted to your finger by the metal. Metal is a conductor of heat energy. "Radiated" heat broils a steak; "conducted" heat frys a steak.

Electric energy can be "radiated" as well as "conducted." A spark coil radiates electric waves. The waves may be "picked up" or received. The first device to receive electric radiant energy was the wireless telegraph.

Waves on water surfaces vary in size. Some are mere ripples, and others may be huge breakers. Electric waves vary in size, also.

A permanent magnet radiates electric energy, but the electricity is similar to the quiet surface of a still pond of water. There are no waves! For radio reception the electric energy must be oscillated or formed into

Inter-office phone set: A switchboard operator is not needed for an office, or business, which has up to twenty desks with phones. The person taking a call pushes but one of the buttons to connect the call to the person being called. The *Hold* button (lower left) is pushed so the receiver may be put back on the hook to allow calls from other lines to come in.

COMMUNICATION

waves; a procedure like that of agitating the water in the still pond.

Radar is the term used to describe the process of "collecting" radio waves which "bounce back" to the station sending out the original waves. The waves do not "bounce back" from all materials. Neither do sound waves. Sound waves bouncing back are called echoes. Returning radio waves are "electric echoes" which may be "received."

The waves of radiant electric energy may be "collected" and "transformed" to flow along conductors or wires. Since the waves "vibrate" through the air, an insulated wire coil attached to an aerial collects the electric energy at the receiving end, just as the secondary wire in the induction coil collects the electricity from the primary circuit. The current collected by the coil alternates within the coil circuit. (Review the conditions under which a current flows in the secondary circuit of the previous chapter.)

You have just learned that *direct current* is necessary to operate a telephone receiver. The current being received by the radio, then, must be changed from alternating current to direct current. One of the devices in a home radio to make this change is a rectifier tube. In the experimental radio you may make here is a piece of galena, a crystal which permits electricity *to travel through it in only one direction!* The galena crystal acts as a "valve," through which the alternating current must pass; and since it allows electricity to flow only one way (just like a one-way street), the current coming out becomes direct current.

Study the diagram of a simple crystal-set circuit, Fig. 138.

In all modern radios, there are various radio tubes and parts. Most all sets are connected to an A.C. current which must be transformed to a lower voltage and changed by a rectifier tube to D.C. The various tubes serve different purposes; some are for increasing the sound volume, others for increasing the range or distance which the set may "reach," etc. In addition, there are devices to aid in tuning and to control the volume and the quality of tone.

You may wish to understand more about radio later on. The control of electrons in vacuum tubes is interesting. Electronic tubes are now used by man for hundreds of purposes other than in radios.

Do not be fooled by a school offering "mail order" courses in electronics. Investigate them thoroughly.

Fig. 138. Crystal radio circuit and finished panel.

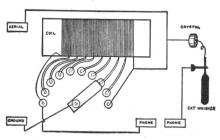

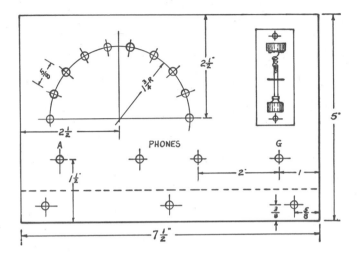

Fig. 139. Panel layout.

JOB 32. Crystal Radio Set

Purpose: To make a small crystal radio set with reception up to 25 miles. See Fig. 138.

Steps:

1. Obtain a headphone set; a galena and catwhisker or a fixed crystal (such as a Germanium crystal); 50 feet of No. 26 single-cotton, enameled wire; one piece of 22-gauge (approximate thickness) copper ½″ x 2½″; one waxed-paper cardboard core; ten brass, ⅜″ No. 1R Dennison two-leg paper fasteners; six ⅜″ 6-32 RH brass machine screws; four binding posts or Fahnestock connectors; one piece of pine ¾″ x 5″ x 7½″; three RHB screws ¾″ No. 5; a 1-inch piece 5⁄16″ D dowel; and one piece ¼″ plywood or masonite 5″ x 7½″.

2. Bevel one edge of the pine board ⅛ inch so that the panel slants backward.

3. Lay out the panel for the holes and drill each one ⅛″ D. See Fig. 139.

4. Push the paper fasteners through from the front, and bend the legs back.

5. Wind the coil by starting one end of the wire through two holes and fastening

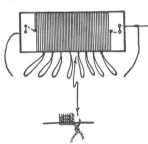

Fig. 140. The coil.

the end secure by twisting. Leave 6 inches for a lead. See Fig. 140.

At each tenth turn, twist a four-inch loop for connections to be made later to the paper-fasteners. After the ninth series of ten turns, cut off the wire to leave 6 inches for a lead; fasten that end as you did the first end.

6. Carefully cut the insulation from the end of each loop and the ends of the leads, and scrape for soldering.

7. Nail or screw the coil to the base after soldering the loops and ends as shown in the diagram.

8. Cut and form the tuning strip and drill a 5⁄32-inch hole through one end. Make a handle from the piece of dowel; drill a hole to fit the end of the copper strip and force into place. See Fig. 141.

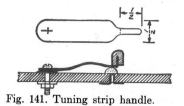

Fig. 141. Tuning strip handle.

9. Install the tuning strip and solder on a lead wire for the ground.
10. Install the detector and binding posts.
11. Complete the remaining connections.
12. Screw the panel assembly to the base.
13. Connect the aerial and ground and the headphones.
14. Test and get approved. A cabinet may be made to fit the panel and base.

THE USE OF ELECTRICITY FOR OTHER MEANS OF COMMUNICATION

There are many other electrical machines and devices which people use to communicate with one another. All of them are developments of the principles of electricity which you have so far learned. A few of these machines are described as follows:

Many newspaper offices and radio news stations use the *teletype*, a typewriter that types the news being sent by a typist who may be hundreds of miles away. Pictures and even messages are now being sent over wires by a process called "facsimile" transmission. Portable telephones like the "walkie-talkie" are well known because of their part in warfare. Television, of course, is being developed for general use.

TELEVISION

With television becoming as popular in every home as is the radio or the family car, a brief description here seems necessary, even though no actual manipulative classwork can be done.

The television camera shoots the scene on one location. What the camera sees and hears is relayed to the transmitting station at a second location. The transmitting station sends out the sight and sound to television receivers wherever they may be.

The camera process is briefly illustrated as follows:

A boy loses two coins in the grass, somewhere in the area between A-B

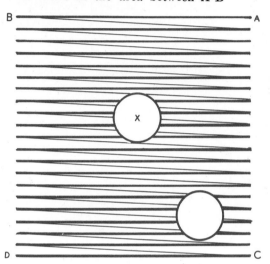

A scanning pattern.

and C-D. He decides to use an orderly way of locating the coins by scanning the area. He starts at "A", scanning to "B", then following along below A-B, etc., as in the above diagram. One coin shows up at "X", and he continues on to "D". The second coin shows up in the lower right.

The TV camera lens is an eye which throws a picture on a small mica plate called the mosaic. Inside the camera and behind the mosaic is an electron gun which scans the mosaic at a rate of more than five hundred lines (as shown above) in less than a third of a second.

Light from the lens forces electrons out of the mosaic. If the lens is focused on the grass (as in the illustration p. 125), the amount of light on the mosaic will be quite different where the two coins are scanned. Thus, the number of electrons will be greater during the scanning only when the scanning paths cross the coins. These variations in electrons are noted by the camera and relayed to a TV transmitter. Here the signal is broadcast to receivers in the area.

The picture tube in the TV receiver consists of a screen, or the glass end of the tube seen by the viewers. This screen fluoresces like a fluorescent light, with each little pin-head-size-area glowing in brightness according to the amount of light in that part of a line scanned by the camera. The picture tube also contains a so-called electron-gun which now works in reverse, but shooting or scanning the large end of the picture tube rather than the small mosaic.

COLOR TV

Behind the glass front of the TV picture tube in a color-TV set is a very finely meshed screen. This is a disc of metal as large as the set opening. In its surface are thousands of small holes, so many and so close together that when held against the light, the disc seems almost invisible.

The holes in the disc, or mask, break incoming TV pictures into small dots upon the inner surface of the tube. These dots are comparable to the ones used for printing pictures in a book. (With a pocket magnifier, examine some of the pictures in this book which have been reproduced from photographs. You will see the tiny dots.)

The surface of the picture tube is coated with tiny specks of different colors. These specks shine, or fluoresce, when hit by a light beam of similar color.

The transmitter camera has three guns for color scanning, and the receiver, likewise, has three. The primary colors of red, blue, and yellow are seen by the eye as *reflected* from surfaces such as paper. But light projected through lenses changes the way these colors appear to our eyes. Try distinguishing the primary colors when wearing yellow sun glasses! (Light which passes through a lens is *refracted* rather than *reflected*.)

The true colors on a color-TV set are tuned by trial and error, with knobs marked on most sets for the colors red or blue or blue-green.

Remote Control by Radio

Control of a machine when the operator is at a distance from that machine is called remote control. Controlling the machine by radio is known as radio control. Many types of missiles are now guided by remote control.

Many model builders control the flight of their model airplanes or the direction and speed of their model boats by remote control. Even lawn-mowers have been so fitted that the operator need but sit on one corner of his lawn and guide his power mower over the grass, around trees, etc.!

In flying a model plane by radio control, the pilot carries a small box which includes the transmitter and aerial. In the plane, a very small receiver receives signals of sufficient

COMMUNICATION

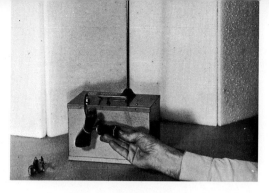

A remote control transmitter. Note escapement in lower left.

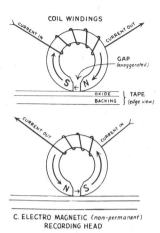

C. ELECTRO MAGNETIC *(non-permanent)* RECORDING HEAD

strength to operate a relay. A battery circuit in the plane from the relay works an escapement, or mechanism, which turns the rudder to the left, to the right, or back to neutral. The plane flies within the control radius of a few blocks, until the engine runs out of fuel. It then glides under the control back to the operator for landing. More elaborate circuits control the motor speed and elevation, so that a flight appears to be controlled by a pilot within the plane itself.

THE TAPE RECORDER

The tape recorder is an instrument of communication for storing sounds for later use. Tape is fed from one reel to a second through a recording head; it is played back in the opposite direction. Special plastic tape is coated with molecules of iron oxide (microscopic particles of steel) which are magnetized into blocks, or areas, each the width of the recording head. The length of the tape varies with the amount required. The magnetizing is done by the recording head, a circular electromagnet with a very narrow gap, as shown at the right.

Sound wave patterns are received by the microphone as currents to be captured by the tape, the currents varying in strength and in polarity. The playback head is very similar to the recorder. As the tape is pulled beneath the head, the magnetic flux of the patterns of magnets on the tape induce electrical impulses in the head. These impulses are sent to the amplifier, to be transformed into sound waves by the speaker system. Keep in mind that the current flow from the secondary coil of an induction coil is varied in strength when the primary coil *receives* a flow of current which is varying in strength, or is reversed. Also, remember that disturbance of the crystals in the microphone caused by the sound waves hitting the crystals causes surges of negative and

The tape recorder.

COURTESY, MINNESOTA MINING CO.

SYMBOLS

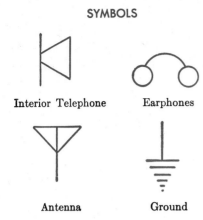

Interior Telephone   Earphones

Antenna   Ground

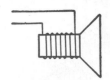

Loudspeaker

positive impulses to the amplifier system.

Tape is erased by recording over a used tape; the polarities of the magnets on the tape are merely rearranged from one pattern to another.

EXTRA JOB *A:* Obtain permission to examine a "hearing aid." Make an enlarged drawing and explain to the class the circuit and the operation of the hearing aid.

EXTRA JOB *B:* Have someone show you how to operate a moving-picture projector. Ask your librarian for some reading material on moving picture projectors. Prepare a demonstration to be given before a class on the use of electricity in the moving-picture projector and how to operate a projector. Form a club of boys interested in projection and offer to help run the school machines during your study periods What part does the photo-electric cell play in the "sound projector"?

EXTRA JOB *C:* Learn about the "electric eye" and prepare a notebook to include clippings, colored drawings, and descriptions of the varied used of the electric eye. Include circuits in your drawings and explain the flow of current.

EXTRA JOB *D:* Make a list of the things to be done in connecting a new radio in the home. Include the installation of the tubes.

EXTRA JOB *E:* Obtain the materials for one-tube radio receiver and wire it to operate.

QUESTIONS

1. Discuss how railroad dispatchers communicate with conductors on moving trains.

2. Describe the telegraph relay circuit.

3. In which instrument, the telephone receiver or the transmitter, is there a permanent magnet? What is its use?

4. Are telephones protected by fuses?

5. What is a diaphragm?

6. Explain how sound is sent over a telephone circuit.

7. What is radiant energy?

8. Do you think that a horseshoe magnet radiates electricity?

9. What is *electronics?* What is *radar?*

10. How many occupations *related* to *communication* at which people earn a living can you list? How does one "enter" these occupations?

LEARN TO SPELL

communication	receiver	telegraph	crystal
relay	vibrate	Morse Code	scanning
telephone	mosaic	transmitter	schematic

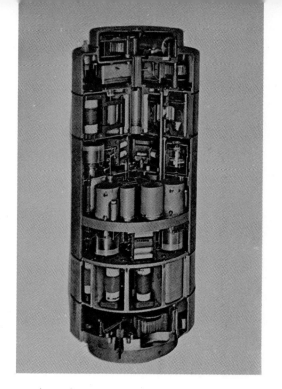

An underseas communication cable amplifier capable of withstanding pressures up to three ton per square inch at depths up to more than two miles · Laying a long distance underground cable.

COURTESY BELL SYSTEMS

128A

Replacing a tape of stored information in a computer.

Bad weather often requires night work by emergency telephone repair crews.

COURTESY BELL SYSTEMS

Erecting a microwave tower for transmitting telephone service over long distances.

COURTESY BELL SYSTEMS

Chapter XI

ELECTRICAL POWER and the Electrical Generator and Motor

> Horsepower — the generator — the electric motor — electricity in the automobile — the storage battery — atomic power — electronics · solar battery.

THE DEVELOPMENT of the electromagnet was the beginning of the use of electricity for power, power to be used by man to run his machines and to do much of his work. In the modern home alone, there are at least a dozen electric motors in machines, to tell the time, to wash and dry clothes, to blow a draft into the furnace or to turn the furnace on or off, to pulverize the garbage so that it may be flushed out the drain, to cool the refrigerator, to mix and stir foods, to clean the rugs, to circulate air in a warm room, to shave whiskers, and so on.

Electric motors in the automobile start the car and circulate the warm air from the heater.

In factories, on the farms, and in the mines, the electric motor does all types of work and often very heavy work!

ELECTRIC POWER

What is power? Power is measured energy, usually represented as the energy measured by the amount of work a horse can do in a given time. One horsepower is equal to the work done by lifting 550 pounds 1 foot in 1 second. Power, then, involves three things: *weight, time,* and *distance.* Motors are rated in horsepower or fractions of horsepower. For instance, the motor used in a typical home washer is $\frac{1}{3}$ horsepower.

For the small electric starter motor in the auto, a very heavy battery is required. Since large motors require a great deal of current, the motors are usually installed where the source outlets of electric power are available; and, since wires are needed for the connections, motors are not operated a long distance from the connection. (See Workbook p. 75.)

ATOMIC POWER

Uranium is a rare element, one type of which is used in making an atomic explosion — one of the greatest forces man has ever created. Splitting an atom of uranium by using a ray changes the uranium into a type of

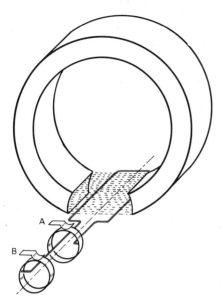

Fig. 142. The simple generator.

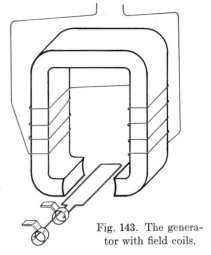

Fig. 143. The generator with field coils.

atom which produces rays for splitting more uranium atoms, in such a fashion as to cause a tremendous, instantaneous release of energy, or *explosion*. This new source of power is so new to man that the problems of harnessing it for practical use have been, and continue to be, as great as the original problem of creating it.

THE GENERATOR

You have learned that the magnetic lines of force radiated about an electromagnet can be picked up by a coil of wire; this is the principle of the induction coil. A coil of wire revolving or turning in a magnetic field also picks up the lines of force. Fig. 142 shows a coil which is rotated in a magnetic field; and because the coil turns, a means of connecting the wire ends of the coil to wires which are not turning is provided by wheels upon which there are two insulated contact surfaces. These wheels are called commutators. Two brushes are installed to rub or contact the two contact surfaces of the commutator. Thus, each time the coil turns and cuts the lines of force, a pulse of current is collected by the coil, to flow out of one of the commutator contacts, to the brush contacts, and then to return through the other brush and commutator contact. See "A" and "B", Fig. 142.

The permanent magnet which "furnishes" the lines of force for the revolving coil may be strengthened by winding coils of wire (called field coils) about the two legs of the magnet and either using some of the current collected from the armature coils or furnishing D.C. supply from another source. Field coils, then, are simply electromagnets, fed by electricity collected from the generator's own poles. (In the alternating-current generator, Fig. 143, the field coils are strengthened by an outside source because the current collected on the rings is alternating current.)

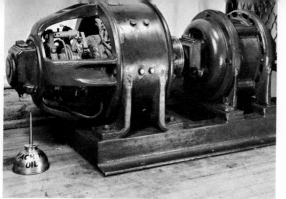

Fig. 144. Motor-generator set.

The polarity or direction of flow depends upon the wiring of the generator and the arrangement of the commutator contacts or segments. There are generators which produce direct current and those which produce alternating current. The flow of current from the D.C. generator is always "direct," in that the current always flows out of one pole and back into the second pole. The flow from an A.C. generator "alternates," the current first flowing out of one pole and back to the second, then out the second and back to the first, etc. The flow changes direction 120 times per second in the 60-cycle generator. Fig. 144 shows a D.C. generator turned by an electric motor. The generator is on the left. Notice the commutator and brushes. In some sections of the country, generators are mounted on towers. A propeller is attached to the shaft of the generator. The wind runs the generator to produce electricity for the owner. See Fig. 145. The electricity generated is "stored" in storage batteries to be used when needed. Fig. 146 shows a generator driven by a gasoline engine.

THE ELECTRIC MOTOR

The principle of the electric motor

Fig. 145. A wind-electric generator.

is the *reverse* of the generator. Trace the circuit in Fig. 148. The current travels from the battery through the field coils, to one of the brushes, which allows the electricity to flow to one segment of the commutator and from thence through the coils (called the armature) and out the other segment and brush.

The armature or rotating part and field magnets become electromagnets, and their poles are attracted to make the armature revolve. The brushes then slide on the simple commutator to break the circuit. The armature poles are not now attracted to the field poles because they have lost their magnetism. The armature rolls on around ("coasting," you may call it), and again the brushes complete the circuit to attract the commutator to the second set of field magnets.

Fig. 146. Generator driven by a gasoline engine.

INDUSTRIAL-ARTS ELECTRICITY

ratings such as the R.P.M.'s, the horsepower, and the voltage. 220 volt A.C. motors are usually three phase, with three lead wires.

JOB 33. A SIMPLE ELECTRIC MOTOR

Purpose: To learn how to make a toy electric motor. Follow the steps and diagrams. Also, see Fig. 149.

Steps:

1. Lay out field pole strip "L" and mark A, B, C, D, E, F, G, and H; center punch and drill holes for ¾" No. 5 RHB screws. See diagram.
2. Bend D, E; then bend C, F. Use round pipe for C, B, and F, G.
3. Form B, A; and G, H.
4. Wind four layers of No. 22 magnet wire around strip at D, E, leaving 6 inches for leads.
5. Mount field pole strip "L" on base "B".
6. Make two bearing brackets "P"; drill running fit for shaft; and drill for ½" No. 4 RHB screws on other end.
7. Cut one piece ⅛" x 4" round drill rod or welding rod for shaft.
8. Make one armature piece 5/16" x 2⅜" round iron "N". Drill hole in center for a drive-fit to shaft "S".
9. Make one dowel spacer "T"; drill and put on shaft "S".
10. Make one dowel for commutator "M"; drill and place on shaft.
11. Cut two pieces thin brass for commutator "U".
12. Wind two layers of No. 22 magnet wire on armature. Starting at middle of wire and center near shaft, wind one end clockwise and the other end counter-clockwise. Solder ends of wire to small brass pieces and tape in place on dowel piece. Make sure opening on commutator is correct in relation to armature position.

Of most importance is the "timing," or the arrangement of the brushes so that they may make and break the contacts on the commutator at the proper time.

You may wish to make a toy motor to help you understand the special importance of timing. Real electric motors have many coils on the armature, each of which is connected to segments on the commutator. See Fig. 147, right. The motor case on the left in Fig. 147 shows the unwound core of the field coils.

There are many ways of winding and connecting armature coils. Motors are made with varied speeds; for example, the vacuum-cleaner motor runs at several thousand R.P.M.'s (revolutions per minute), and the usual washing machine-motor runs at about 1725 R.P.M. Examine the name-plate on several motors to learn the various

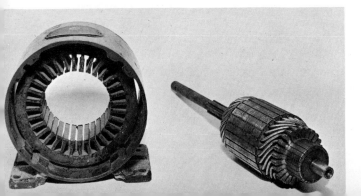

Fig. 147. Parts of an electric motor.

POWER—THE GENERATOR AND MOTOR

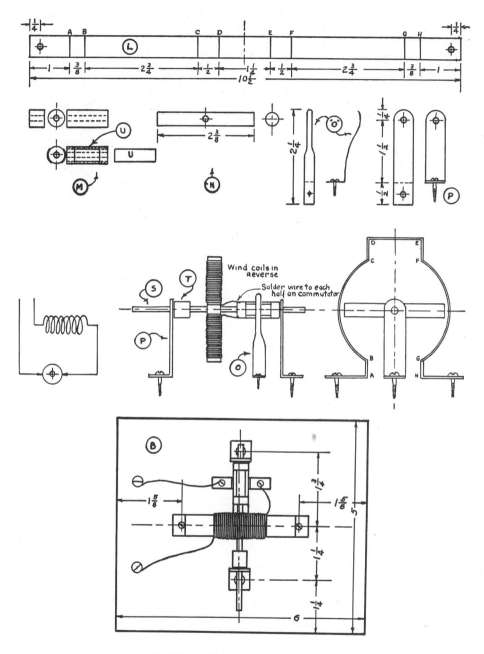

Fig. 148. Simple electric motor.

13. Assemble armature and place in brackets under field strip.

14. Make two brushes, ¼" x 2½" spring brass "O".

15. Mount brushes as shown.

16. Connect wires as shown in diagram (a series hookup).

17. Operate on 6 volts D.C. or slightly higher A.C.

Electric Motor

Bill of Material:

1 Base ¾" x 5" x 6" "B"
1 Field pole strip ½" x 10½" 18 Ga. iron "L"
2 Bearing brackets ½" x 2¼" 18 Ga. iron "P"
1 Shaft ⅛" x 4" "S"
1 Spacer ⅜" dowel ⅜" long "T"
1 Commutator ⅜" dowel 1" long "M"
2 Brushes ¼" x 2¼" — 28 Ga. spring brass "O"
1 Armature ⁵⁄₁₆" x 2⅜" round soft iron "N"
4 Screws RHB ¾ No. 5 (bearing brackets and field)
2 Screws RHB ½ No. 4 (for the brushes)
1 Magnet wire No. 22 Ga. ECC (four layers on field pole strip)
1 Magnet wire No. 22 Ga. ECC (two layers, each end of armature)
2 Brass ⁵⁄₁₆" x 1 — 28 Ga. (commutator assembly) "U"
Tape

Induction Motor

The induction motor shown in Fig. 150 is an electric motor with no armature coils, no brushes, and no commutator. The rotating part is made of laminated segments arranged in a spiral row around the outside. Space does not permit complete explanation here; but, in a brief description, the field coils so affect the polarity of the segments as to automatically attract and repel parts of the segments and thus make the motor run. If you are interested, ask your school or community librarian for a book on electric motors and study the induction motor.

Synchronous Motor

The electric-clock motor is a simple type of synchronous motor which operates only on alternating current. The toy model on the far right, Fig. 149, shows such a motor. Recall that the polarity of 60-cycle alternating current changes 120 times per second. It is this fact upon which the operation of the motor is based. Each arm of

Fig. 149. Toy electric motors.

POWER—THE GENERATOR AND MOTOR

Boy checking his synchronous motor.

the *rotor* is attracted by the *stator* at each change of polarity in the current, and current changes keep the time accurate.

The Care of Motors

Few motors in the typical home need more than a periodic inspection to check for over-heating, dust, or bearings which need oil. The bearings are the supports for the two ends of the armature shaft which permit the shaft to turn. See Fig. 150, and notice the bearings and felt pads to retain the oil. An oil-hole screw for each bearing is also shown. When bearings become badly worn, the armature may strike the field magnets and cause a breakdown.

Many recently built motors have sealed bearings, which means that the bearings are so constructed as to never need oiling. But look for an oil hole on each bearing of the motors in your home. Directions are furnished with many new appliances which tell how often to lubricate the bearings and how much grease or how many drops of oil to apply.

Too much oil may cause as much trouble as no oil. Oil may work into the commutator and brushes and interfere with the contact. Too much oil may soften the insulation on the wires and also collect dust to prevent air circulation, thus causing the motor to heat.

In motors where heavy grease or hard oil is used for the bearings, grease cups are provided which must be "turned down" occasionally and refilled with a good grade of hard motor oil.

The brushes are the contacts which slide on the commutator segments. Carbon blocks held by springs against the commutator are typical brushes. When the carbon wears out, new ones must be installed. Installing brushes on many motors is not a difficult job.

The carbon from the brushes often fills in the space between the commutator segments, seen in Fig. 148. Repairmen clean out the carbon and turn

Fig. 150. The induction motor.
COURTESY, A. G. REDMOND CO.

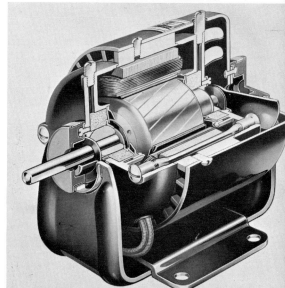

down the commutator on a machine lathe when this condition arises.

ELECTRICITY IN THE AUTOMOBILE

As you probably know, it is an electric spark between two contacts in the spark plugs of an automobile engine which explodes the gasoline and air mixture to drive the piston down and run the motor. The *ignition system* refers to the electrical circuits which provide these sparks.

In addition to the use of electricity for the ignition system, electricity serves many other purposes in the automobile, such as for the lights, for the electric clock, for the cigar lighter, for cranking the motor, etc.

The source of electricity in the automobile is the generator, driven by the running engine. Electricity from the generator is stored in the storage battery to be used as needed. Fig. 151 shows the ignition circuit in the typical auto. Notice that one pole of the battery is grounded.

The Ignition Circuit

The distributor is a timing device which makes and breaks the contacts in the primary circuit of the induction coil to send a strong spark to the spark plugs at the exact time it is needed. Two contact points make and break

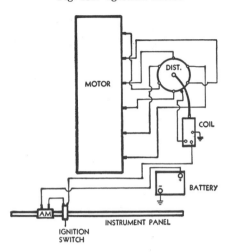

Fig. 151. Ignition circuit.

the primary circuit in the distributor. Heat is generated by this arcing, so tungsten is used for the metal in the points, because it will not melt except at a very high temperature. Even then, points need to be filed or changed completely from time to time.

The condenser is wired in parallel with the primary circuit of the coil. The condenser is made of insulated layers of tin foil wound in a bundle for compactness and usually located inside the distributor. When the circuit is *open*, the electricity is stored up on the surface of the foil; and when the circuit is closed the current rushes with great force through the coil to create a high voltage impulse in the

Fig. 152. The spark plug, wire brush, and gap gauge.

POWER—THE GENERATOR AND MOTOR

secondary coil. The points are opened and closed by a cam, an off-center wheel.

The space between the two contacts in the spark plug is called the gap. For the motor to run properly, this gap must be set accurately.

JOB 34. CLEANING AND SPACING THE GAP ON A SPARK PLUG

Purpose: To learn how to inspect the porcelain jacket and to clean and adjust the gap on a spark plug.

Steps:

1. Obtain several odd spark plugs, a thickness gauge, and a wire brush. See Fig. 152.
2. Brush the inside part of the plug and clean the porcelain with a cloth. Look for cracks in the porcelain.
3. Measure the gap with the gauge and tap the ground electrode lightly on a hard surface until the .032 inch gauge blade fits snugly between the two contacts.

The Lighting Circuit

Fig. 153 shows the lighting circuit on cars with only two headlights. Cars with four headlamps are wired similarly. In each pair of lights, both lights are grounded, and the two control wires running from the foot switch each go to a different light. Why? How many lights can you count on today's cars?

The Fuses

Cartridge fuses are used in the lighting circuit and are usually placed behind the instrument panel. See Fig. 153. For night driving, each driver should know the location of the fuses in his car and should keep extra fuses in a convenient place. Then should a fuse blow while he is on the highway at night and has no access to a flashlight, he may replace the fuse in a minimum of time. A car on the highway at night without lights invites disaster.

The Ammeter

When the motor is running, the ammeter shows the amperage which the generator is producing. Cars without ammeters have a discharge warning light. Leaving the ignition switch on for any length of time will "run

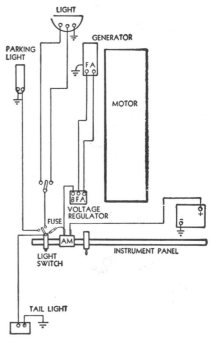

Fig. 153. Simplified lighting circuit and generator circuit.

the battery down," or, in other words, use all the amperage stored in the battery.

Trace the circuits of several of the lights through the various types of switches. See Fig. 153. Again, notice that nearly all of the lights have one contact grounded. If you think the circuits in the automobile are complicated, find a plan of the circuits in a modern airplane! A good auto mechanic must know how to trace all the circuits in all makes of cars when locating trouble. However, he may be furnished the wiring plan, to save time.

There are many men who are specialists in the field of automotive electricity and do nothing else but repair the "nervous system" of automobiles.

THE STARTING CIRCUIT

The starting circuit is fairly simple. Trace the circuit in Fig. 154. You undoubtedly have noticed that the lead wire to the storage battery is very heavy. In fact, it is called a cable instead of a wire. Since the motor must be fairly strong to turn over a cold engine, especially in cold weather, the starting motor draws very heavily on the battery. A large conductor is necessary to carry the high amperage required under those circumstances. Likewise, the starting switch must carry a heavy load.

THE STORAGE BATTERY

Car storage batteries contain two-volt cells wired in series, three to the 6-volt battery and six to the 12-volt. A fully charged battery has more than a hundred amperes for immediate use. To prevent overcharging of a battery, a cutout in the generator shuts off the supply of amperes going to the battery.

Briefly explained, the storage battery is made of plates separated by insulators and submerged in an acid solution. In charging a battery, the electricity is changed to chemical energy within the plates. In using electricity from the battery, it is changed back again.

The liquid in a battery is sulphuric acid and water. The water evaporates in time and must be replaced to prevent warping of the insulators between the plates and thus ruining the battery.

Liquids vary in "density"; for example, molasses is very dense, and alcohol is the opposite. A cork will not sink as far in molasses as it will in alcohol. The liquid in a storage battery varies in density as the amperage in the plates varies! It is said that in charging a battery, the electricity forces the acid out of the plates and into the solution or liquid, making the liquid more "dense." The opposite action occurs when a battery is being used. A glass-sealed tube will float higher in the liquid of a fully charged

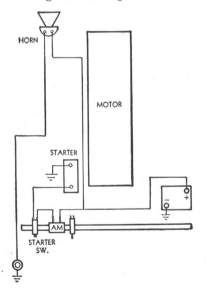

Fig. 154. Starting circuit.

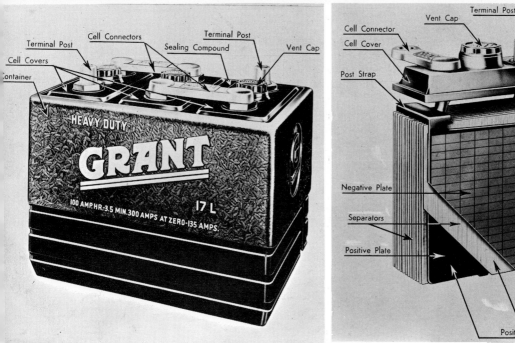

COURTESY, GRANT STORAGE BATTERY CO., MINNEAPOLIS

Fig. 155. Cross-section of storage battery.

battery than in the liquid of a "dead" battery.

The hydrometer shown in Fig. 156 is such a sealed glass tube, with a weight in one end to keep it upright and a roll of printed paper in the other end for reading the density (called specific gravity). The tube floats in a syringe.

JOB 35: TESTING A BATTERY WITH A HYDROMETER

Purpose: To learn how to care for the car storage battery.

Steps:

1. Inspect the cable connections on a storage battery in a car. The greenish, corroded deposit on the poles should be scraped off and the terminals wiped clean and dry. Cover the terminals and connections with heavy grease. Wipe off the top of the battery. CAUTION: The liquid in a storage battery is a strong acid solution (called the electrolyte). The acid is dangerous!

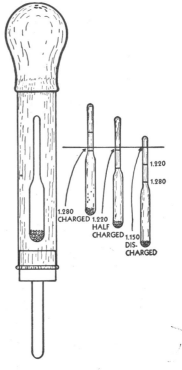

Fig. 156. The hydrometer and float.

2. Remove the vent plugs, being careful not to drop dirt or any other foreign matter inside the battery.

3. Obtain a hydrometer, insert the tube in one cell, and squeeze the bulb.

4. Release the pressure to draw up enough liquid to free the float from touching either the top, bottom, or sides of the hydrometer. *Keep the end of the tube inside the cell.*

5. Read the charge on the float. Make sure the float is not touching any part of the hydrometer but floating free. Take the reading at the liquid line on the float. See Fig. 156.

6. Force the liquid back into the cell. If you lay the hydrometer down, make sure that no drops are deposited for someone to touch.

7. Test each cell, recording all three readings on a piece of paper.

8. Add distilled or approved tap water to the cells to fill them to a fraction of an inch *below* the lower end of the vent plug opening.

9. Wipe off the hydrometer tube with an old cloth and return to safe storage.

10. Find the true reading of each of the three cells by this method:

   a. Obtain the thermometer reading of the temperature of the liquid by inserting a thermometer for five minutes, and call this temperature reading "T". (If battery has been for four hours or more in the same room where a wall thermometer is located, the wall thermometer reading may be used.)

   b. If the temperature of "T" is above or below 70° (Fahrenheit), subtract the difference and divide the answer by 3,000. Call this answer "X".

   c. If the temperature is *over* 70°, *add* "X" to the hydrometer reading; if the temperature is below 70°, *subtract* "X" from the hydrometer reading.

   d. A fully charged battery should have a
   "corrected reading" of about 1.280
   A half charged battery   1.220
   A discharged battery   1.150

11. To charge a battery on a battery charger, follow the directions furnished by the maker of the charger. Connect the positive pole of the charger to the positive pole of the battery and connect the two negatives together. *Do not overcharge.*

Always keep the electrolyte level above the plates by adding water. Do not overfill a battery, as the acid will eat at the strap fasteners or cable insulation or will corrode the terminals.

The numbers on a hydrometer float compare the "density" of the liquid to the "density" of water. This relationship is called *specific gravity*. A liquid which is twice as heavy or dense as water has a specific gravity of 2.000. The heavier the liquid the greater the specific gravity. The specific gravity of the electrolyte (the storage-battery liquid) in a storage battery varies from 1.150 to 1.280.

THE GLOW PLUG ENGINE

Most model airplane engines use a glow plug for igniting the compressed fuel-air mixture in the cylinder. The glow plug is a spark plug with a built-in heating-element coil. A dry cell is required to start the motor, but once the engine is running smoothly, the battery is disconnected. The engine then continues to run, because the heat from the explosions keeps the coil glowing hot. Special fuel is mixed for glow plug operations, a type which explodes at a fixed temperature and at a given compression.

The typical power lawn mower, and snowblower, have gas engines and are equipped with a small spark plug, similar to the spark plug used in cars. However, they do not require batteries, such as those needed for a car or model airplane engine. Outboard motors for boats (except those with electric starters) also do not need batteries. A magneto type of generator

POWER—THE GENERATOR AND MOTOR

A glow plug.

furnishes the electric charge for the spark plug after the starting rope is pulled.

RAILROADS AND ELECTRICITY

For rolling stock, the railroads use steam power, diesel-electric power, and electric power from electric motors which obtain current through a third rail or overhead cable.

The diesel-electric locomotive is replacing many steam-driven locomotives. The axle of each wheel of the diesel-electric is the extended shaft of the rotor of an electric motor. The motors receive current from a large generator powered by an oil-burning diesel engine. A large locomotive may consist of two "A" units, one in front and one facing the rear with one or more "B" units flat on both ends in between the two "A" units.

The uses of electricity in railroading are countless. There are devices to increase the safety and comfort in traveling, as well as those which are used in operating and maintaining the roadways.

ATOMIC POWER

In this era of the atom, man has learned how to obtain power from the tiny atom in amounts varying from the weak signals of radio waves to the tremendous force of an atomic explosion. He has learned how to parcel this power in an A-Bomb or H-Bomb in smaller packages to be used where less energy is required. He has learned how to use the radio waves to guide rockets through the earth's atmosphere into orbit around the earth or sun. *Power,* controlled by the use of a small glass bulb—the vacuum tube—is one of man's recent marvels of accomplishment.

ELECTRONICS

Electronics is that branch of electricity in which vacuum tubes and transistors play the leading roles. Take a good look at the period at the end of this sentence. It contains millions of atoms of the chemicals in printer's ink! Each atom is a tiny solar system with a small, hard center of positive charge; and revolving around it at unimaginable speed are the negatively charged electrons, moving so fast as to give it a shell, much as the revolving propeller of an airplane appears to have a circular piece of glass around its hub.

A glow plug engine mounted on a test stand.

Under certain conditions, one or more of the electrons may escape from its proton (center). You walk across a wool rug; your shoes knock some electrons loose; they travel up through your body; you bend over a drinking fountain; they find a good escape route from the end of your nose to the bubbling water and make a wild jump to escape from you and return to "equilibrium." Your nose knows that something happened.

Heat a wire electrically in a glass sealed tube from which the air has been removed; and the electrons come out of the metal wire like millions of bees from a comb of wild honey that has been pushed from an old log by the nose of a bear.

But a bee is a huge beast compared to an electron! It is estimated that it takes 6,000,000,000,000,000,000 (six million trillion) electrons a second to keep a 100-watt bulb burning.

The Vacuum Tube

The well-known vacuum tube is, then, a device for freeing electrons from their atoms and making these electrons do work. There are tubes where the electrons are forced out of the atoms in the metal wire by heat; tubes which are not vacuums but filled with certain gases and where the electrons forced from the heated wire "knock off" more electrons from the gas atoms to further increase the electron flow; tubes in which there is no *heated* wire but in which light beams from outside the tube "smoke out," so to speak, the electrons from the atoms in a chemically coated plate in the tube, etc.

See the diagram above of a simple vacuum tube. "AB" is a wire filament running from a power source (not shown) into and out of a vacuum tube. The "free" electrons within race madly around, hitting the walls of the tube or anything in their path. "X" is a metal plate with a sealed connecting wire extending out of the tube. The electrons, negative in nature, hit the plate and give it a negative charge.

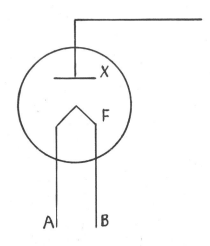

Suppose another circuit is added as below. Here the battery (2) is connected to a meter to register the voltage. The wires (5) and (6) are connected to the filament and plate. When the switch (S) is closed, the electrons flow from the plate (X), along wire (5) to wire (3), through the meter, along wire (6), into the tube via the filament and across the gap to the plate, to complete the circuit and increase the meter reading (provided the volume of electrons from (F) is greater than those flowing from the battery).

Now suppose that a wire mesh screen is placed between the filament (also called the heater) and the plate, and that the screen or grid has a connection outside the tube, see page 143.

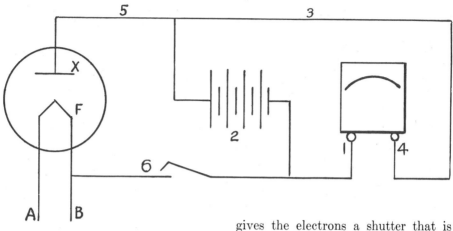

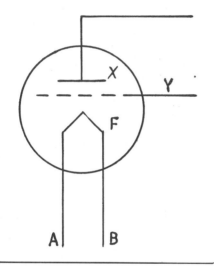

The electrons are now forced to go through the screen much as sunlight flows through an open window-shutter. If a positive charge is applied to the grid (Y), the electrons cannot get through. The window-shutter is closed, no sunlight can enter.

As you recall from your previous study of induction, the radio aerial picks up energy from radio wave signals, alternating plus and minus charges. Connecting the aerial to (Y) gives the electrons a shutter that is being opened or closed to a degree that varies with the strength of the signals. Thus, the varying current from (X) can be changed to sound waves by use of a telephone receiver —the principle of the detector tube. If another plate, called a cathode, is placed near the heater, additional electrons will flow from this plate— to increase the strength of flow: a type of amplifying tube. The cathode may be heated by an A.C. current on the filament, to thus eliminate the need for batteries.

There are many, many types of vacuum tubes designed for many different purposes. Mention has been made of the transistor, the small device that functions like some tubes. Can you find out, in this great field of electronics, how the vacuum tube or transistor is used to:

1. Steer a model plane or a guided missile.
2. Charge a car battery.
3. Control the voltage on a generator.
4. Protect a power line from lightning.
5. Set off an H-Bomb (well you'll have to guess here).

6. Open a garage door, long before you get to the garage.
7. Turn on a drinking fountain.
8. Measure the amount of fat on a hog.
9. Guide a submarine under the polar ice.
10. Dim the lights on an approaching car at night.
11. Chart the strength of your heart beat.
12. Bring in an airplane without touching the controls.
13. Find cracks in metal.
14. Sort out beans.
15. Find a file in a prisoner's gift cake.
16. Check the compression in your car's engine.
17. Bring in Hi Fi music.
18. Baby sit.
19. Tune a trombone.
20. Find a buried treasure.
21. Make a tape recording.
22. Determine the age of an old bone.
23. Make organ music.
24. See the crack in a broken leg.
25. Sort boxcars in a switch yard.
26. Automatically set the shutters on a camera.
27. Clean dust and smoke from the air.
28. Detect an enemy long range missile.
29. Measure the dangerous atomic radiation in the atmosphere.
30. Count sheep.

A transistor. "This transistor is used as an amplifier in a telephone system. Inside the metal shell are two hair-thin wires resting close together on one side of a small wafer of germanium, the opposite side of which is attached to the base. Unlike the vacuum tube there is no glass envelope, no heating element to cause warm-up delay and, of course, no vacuum."

COURTESY, WESTERN ELECTRIC

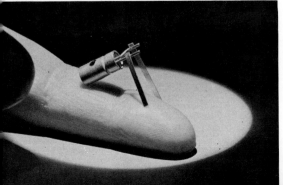

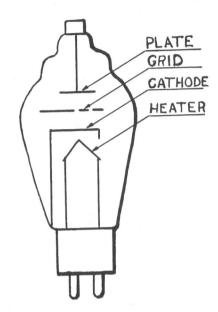

Amplifying

Study the circuit, page 145, with its triode tube. The triode has three prongs in addition to the two heater leads (pronounced leeds). Note the cathode over the heater, the plate, and the grid. Here the triode is used to enlarge or *amplify* the current, or rather the surges of current fed to it by the transformer at the input end. The electric waves have been received, or rectified, by the unit leading to the input. They are being enlarged here and passed on to the next unit at the output. The signals enter at "A", through the primary of the transformer at "1". Trace the secondary circuit from "3" to "4" to the cathode; to the grid; through the secondary and return to the power at "2". What is happening at the grid to affect the circuit to transformer "B"? Trace this second circuit from "6" through transformer "7"; to the plate; to the

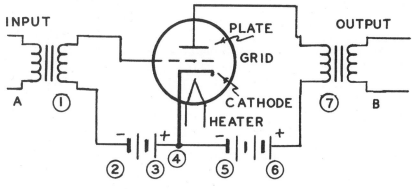

One tube amplifier.

cathode, to "4" and then to "5". The power source to the heater is omitted to simplify the drawing.

Transistors

The transistor has gradually taken over the work done by vacuum tubes in many electronic circuits; this is because it has many advantages over the tubes. Transistors are much smaller than tubes and require much less current to operate. For example, an early computer equipped with tubes used as much space as that of a classroom and had to have special cooling equipment to keep the room from being set on fire. The same machine, redesigned, using transistors instead of tubes, required space about equal to that of a grand piano, and cooling equipment was not needed!

At right is an amplifying circuit using a transistor rather than a tube. See how it resembles the schematic above. Read and think about the following on the transistor:

Two elements, silicon and germanium, are used in their crystal form. All crystals have very interesting arrangements of their molecules, and these two elements are unique in that their atoms are paired. That is, two atoms of silicon are stationed close together, with their electrons orbiting in pairs. Both silicon and germanium are insulators, but when they contain a trace of arsenic, or boron, they become semi-conductors.

When arsenic is added, one additional electron is added to the orbiting pairs of electrons, allowing one electron to be pushed out of orbit and into a circuit. If boron is fused instead of arsenic, an opposite effect results, because boron is known as a 3-valance bond (arsenic is a 5-valance electron bond element). Thus the paired electrons orbiting around the nuclei of their atoms are short-changed by their one electron. A *hole*,

One transistor amplifier.

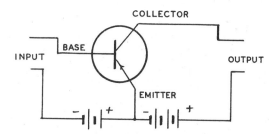

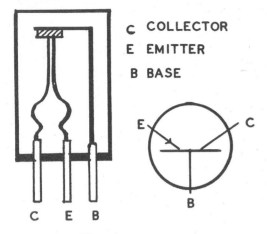

The point contact transistor.

1. A base (comparable to the grid of the vacuum tube).
2. A collector (comparable to the plate).
3. An emitter (comparable to the cathode).

Transistors are manufactured in two ways, by the point contact and the junction-type methods. The point-contact transistor is built with two wires, called cat whiskers, which are firmly fixed. One end of each wire is secured at the base, and the other end of each wire is attached to the crystal. One wire is called the emitter; the other, the collector. Leads come from each. The base of the transistor is the contact to the crystal. See drawing at left.

The *junction-type transistor* is a boron impurity (P-type), with one slice with arsenic (N-type) sandwiched between the two. The base connection is made to the N-type slice. One of the P-type slices becomes the emitter section, while the other P-type becomes the collector terminal. This particular arrangement of the three segments of crystals is called the P-N-P transistor. The N-P-N transistor is another type, which has two N-type crystals with one P-type.

The junction of the lead with the crystal is made with a spot of indium, an easily fused element. The circles enclosing the plus signs represent the positive *holes*, which, when forced by a battery flow, attract the electrons and build up the imput surge of current to the output. Then signals received through the input circuit, are duplicated in the output circuit but in a greatly magnified manner. Transistors, to furnish the same amperage as

or positive charged area, is thus created. These positively charged holes throughout the crystals work with an electric current to force the electrons to move in a controlled direction, thus increasing the flow.

As a result, germanium and silicon are insulators until impurities such as arsenic or boron are added, to make them into semi-conductors capable of permitting a controlled flow of electrons.

The addition of arsenic changes the crystal to the negative, or N-type crystal. However, the addition of boron changes the crystal to the positive or P-type crystal. Each transistor has the following three parts:

Heavy-duty transistor of the P-N-P germanium variety with an output of 65 amperes.

COURTESY, HONEYWELL

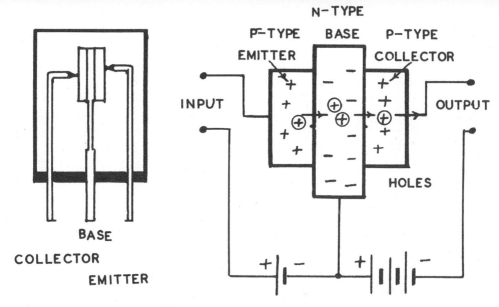

The junction type transistor.

vacuum tubes, need but a thousandth of the power from the battery source to obtain the same result.

Understanding the circuits in the radio receiver furnishes a broad basis for understanding similar circuits in hundreds of electronic devices. The typical home radio contains the following circuits: an antenna system, a detection system, an audio circuit, an automatic volume control, and a power supply system.

On the professional level, the electronics engineer knows how to design circuits to accomplish a given purpose. Technicians in electronic fields operate, maintain, and repair electronic equipment. Few technicians know the complete theory of operation of each circuit, but all are capable of reading the circuits and knowing how to analyze, to locate, and to correct the troubles.

The hand tools most often used by electronic servicemen are the screwdriver, the soldering iron, and pliers—along with a few test meters. Prepared kits of electronic parts to be assembled into such items as radios, amplifiers, meters, and the like, are excellent for the learner as a foundation for becoming an engineer or a technician. You may wish to start early on such a career by using your spare time to build such an apparatus.

It is estimated that approximately half of the defective radios seen by repairmen need only replacement of one or two burned-out tubes. Almost every neighborhood has a store which furnishes a self-service tube tester, so that customers can test their own radio or TV tubes—except the picture tube. Tubes must be plainly marked, so that they can be returned to their correct holders. All sets have the tube charts posted in the back, and all tubes are numbered. Many tubes have a substitute type which will be just as effective in the set, but will carry a different number. Therefore, it is wise to make a separate tube chart, showing the exact number on the tube taken from each holder. CAUTION: Always disconnect a radio, or a TV set, from the house circuit before removing the back cover. Allow a TV set an hour after it is turned off before you remove the back cover. One can still receive a dangerous shock for some time after a TV set is disconnected. However, a serviceman will gladly check your tubes in his shop without charge, and he knows how to make a more reliable check than the amateur.

Almost all electronic circuits use resistors, capacitors, and induction coils. A thorough understanding of these terms requires more advanced study of electronics than is possible in this text. Very briefly, they are described below:

Resistors are included in circuits to control the flow of electricity. Fixed resistors, obtainable in values of resistance from a few ohms to hundreds of thousands of ohms, are small carbon cylinders coated with plastic. A wire lead is attached to each end of the resistor, and the resistance of each is marked by coded bands. A potentiometer, so familiar as the volume control on many radios, is a *variable resistor*. The maximum resistance is stamped on the case.

Capacitors (also called condensers) are a storing device for electrons.

Variable capacitors, usually used for tuning stations in or out, are made of sets of plates. One set is fixed permanently in a frame, and the other is on an axis to be rotated through the spaces of the fixed plate. The typical fixed capacitor is made of two thin aluminum foil sheets separated by insulation paper, rolled into a tight cylinder and waterproofed. One lead is provided to each sheet of foil. See illustration below.

The quantity of electrons that condensers can hold (called capacitance) is measured in farads, or rather in fractions of one farad. A microfarad (mfd) is one millionth of a farad. A micro-microfarad (mmfd) is a millionth of a microfarad.

Capacitors have too many uses to be listed here. In radio, they block a direct current from entering a circuit, but at the same time, they allow an alternating current to go through. The capacitor in an automobile (called a condenser) permits electrons to pile up on one of the plates so that the spark plug receives a very hot spark the moment a distributor rotor wipes the contact.

Induction coils have been described in a previous chapter, but not as they apply to electronics. Inductance slows down any change of current. Inductance makes radio tuning possible.

The following schematic of a simple transistor receiver includes parts available at most electronics supply houses. First, list the parts needed, and obtain them. Remember, where tap joints are shown, use the point connectors and solder to the lugs. Buy the small 1½ volt cells.

Capacitors (three on left); resistors (three on the right).

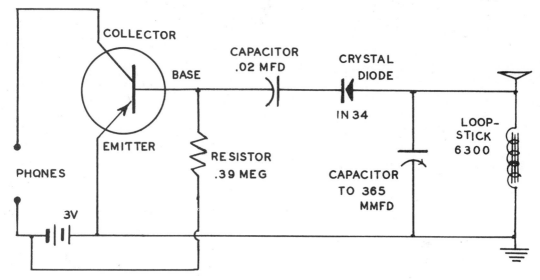

One transistor receiver.

New Age of Power

One of Einstein's great contributions to the new age is his formula $E = MC_2$ where E means energy; M stands for the mass or weight of a substance; and C_2 indicates the speed of light multiplied by itself. The potential energy in a substance is equal to the weight of it multiplied twice by the speed of light. The formula inspired other scientists and their combined efforts developed the splitting of the atom, the chain reaction, the A-Bomb and the H-Bomb. The results of their efforts ended a war, changed the methods of defense of nations, and made the outlook of any future major war so terrible that it is hoped no nation will dare to start another. But man immediately started to harness this great power for peace-

Atomic power plant.

COURTESY, NORTHERN STATES POWER CO.

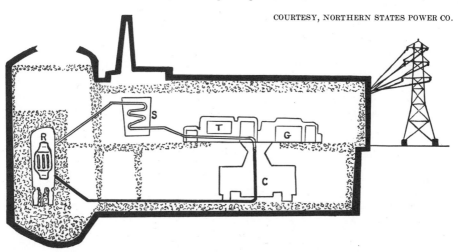

ful purposes. Here are a few; but even before another one of those printer's periods can be placed at the end of this sentence, other developments in the field will be announced.

The Atomic Power Plant

Many electric power plants using atomic energy for power are now in operation. The atomic engine consists of a heavily lead-shielded reactor (See "R" in the following diagram) in which the fuel, uranium, is used to generate heat. One pound of uranium yields the heat energy of 1,360 TONS of coal. Water is made into steam in the reactor (R); the steam is super-heated with conventional fuels (S); the steam drives a turbine (T); the turbine powers an electric generator (G); and the electric current produced is distributed over power lines to the customers. The condenser (C) changes the steam back to water.

Atomic engines are also used in submarines, ships, and it is said, large airplanes. Scientists are experimenting with the use of atomic power for changing the salt water from the ocean to fresh water for irrigation and drinking purposes. Their hope is to develop a method which will make fresh water available at low cost.

Rocket Power

Electricity is used with rockets mainly for firing and controlling the flight. Both solid and liquid fuels are used for power. The purpose of the rocket is to push a self contained device located in the nose-cone of the rocket through the earth's atmosphere and into orbit around the earth, or the moon, or the sun. The device in the nose may be an atomic war head, if used in a war, or it may be a device that contains electrical equipment for obtaining information, measuring it, and broadcasting it back to earth. Such satellites help meteorologists in weather forecasting, astronomers in studying the earth, its atmosphere and neighbors, scientists concerned with national defense, etc.

Solar Power

The average home uses from one hundred to two hundred kilowatt-hours per month. The sun supplies the earth with a thousand trillion kilowatt-hours of energy *each day*. The greatest source of energy of all is from our sun, but man has been unable to harness this power to any great extent. He has designed a few homes which make maximum use of the sun's rays for heating purposes; he has experimented with lenses and reflectors to concentrate the heat waves; but all in all, his efforts have not been too successful. The solar battery, developed by the Bell Laboratories, however, seem to have possibilities for obtaining power for certain uses.

The sun gives off energy in the form of heat rays and light waves. The unit of measure used to describe the energy of light waves is called the photon. The principle of the solar cell which uses the photon, is illustrated on preceding page.

Silicon is an element commonly found in sand. Glass is made from sand along with other chemicals. When the sand is melted, it forms, when cooled, a crystal mass, much as sugar does when it is melted and allowed to cool undisturbed. The solar cell is a thin silicon crystal wafer

POWER—THE GENERATOR AND MOTOR

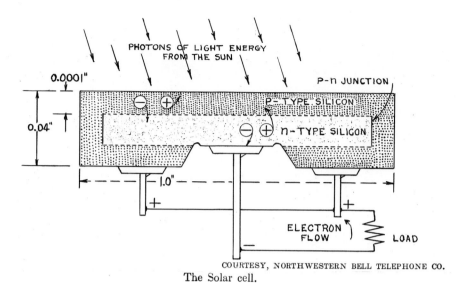

COURTESY, NORTHWESTERN BELL TELEPHONE CO.
The Solar cell.

known technically, positive "holes" in the boron layer. This causes a flow of electrons or current into the circuit shown in the diagram. Each wafer produces a small amount of energy but by combining many of them into a battery the accumulation is sufficient to power a telephone. To be sure, power is available only during the sunlight hours. An even flow is made possible by using storage batteries to carry on when the sun is behind heavy clouds or when the sun has set.

The photograph below shows a Bell Telephone lecturer explaining the Solar Cell. He exhibits a picture of a Solar Battery located on a telephone pole, which is supplying power for a rural telephone system. On the table are other electronic devices. Can you identify them?

LASERS

You will recall that electrons may be forced from one orbit to another and when this takes place, a tiny photon of light is emitted. Again,

Electronic devices.
COURTESY, BELL TELEPHONE

A laser scientist. COURTESY, MINNEAPOLIS HONEYWELL

you learned about N-type and P-type semiconductors, using the elements of silicon and germanium. Hundreds of research teams are now experimenting with a new type of light source called *lasers*, (the acrostic formed from the words light amplification by stimulated emission of radiation). One group has added impurities to the element gallium to make a P-type and an N-type semiconductor. A block of each of these metallic semiconductors are joined to form a *junction*. The two ends of the block are very highly polished. When a direct current is connected to the two blocks, strange things happen within the block and specifically at the junction lines of the polished ends. The atoms within the two blocks are so stimulated by the current that electrons are forced from one orbit to another, emitting photons of light. In so doing, these photons knock other electrons from orbit which in turn emit photons to knock other electrons out, etc., etc.

POWER—THE GENERATOR AND MOTOR

Many of these rays, on hitting the polished ends of the metal, are reflected back to do more of this displacement of electrons. The interesting phenomenon that occurs is that light rays shoot from the line of junction at the polished ends, rays which have but one wave length and a light which is new to our earth. Laser beams have pierced holes in razor blades and diamonds, made intricate welds, and carried signals. But the practical uses to be made of this new power are yet to come.

Ruby crystal rods are also used for producing lasers. Ruby is a pink colored aluminum oxide in crystal form, to which very small amounts of chromium have been added in its formation. The ends of the crystal rod are highly polished, acting as reflecting mirrors within the rod. Light is thrown onto the outside of the rod, forcing the electrons within the rod out of their orbits. Again, this movement of electrons from their orbits to other orbits produces the additional flow of electrons out of orbit, etc. In addition, the reflections from the mirrored ends causes still more electrons to be forced out of orbit, until it becomes a veritable avalanche of light energy.

EXTRA JOB *A:* Obtain a ¼ H.P. motor such as one from a washing machine. Disassemble and examine the various parts. Assemble.

EXTRA JOB *B:* Ask the garage owner for a defective solenoid starting switch discarded from a Ford or another make of car which uses it. Disassemble and make a drawing of the circuit.

EXTRA JOB *C:* See if your librarian has a book on locomotives, and study the Diesel-electric locomotive.

EXTRA JOB *D:* Make a circuit drawing of a toy electric train. Show and describe the various types of switches.

EXTRA JOB *E:* Make a drawing of the ignition circuit on a model gas plane motor.

EXTRA JOB *F:* Obtain from a car dealer a *manual of instruction* for a car which has left and right turn signal lights. Study the operation of the signal and see if you can draw the circuit.

EXTRA JOB *G:* Prepare a five-minute talk on atomic power. Your librarian will help you find the source material.

EXTRA JOB *H:* Make a list of 25 different uses of electricity in railroading. Describe each to the class.

EXTRA JOB *I:* Bring a toy train locomotive to class and show how the reversing unit changes the direction of the motor.

EXTRA JOB *J:* Bring to class a small HO layout and describe the following to the class:
1. What HO gauge means.
2. How locomotive and car kits are assembled.
3. How track is built.
4. Why D. C. current is used.
5. Explain how turnouts are wired, and how engines are controlled as to speed and direction.

Each block is insulated from the two adjoining blocks.

EXTRA JOB *K:* Obtain a small ½A model airplane glow-plug engine, and demonstrate to the class how to start and adjust it for running.

QUESTIONS

1. Make a list of the uses of electric motors which you know are being used in a local grocery or garage.

2. Make a list of the uses of electric motors used on a large farm in addition to the uses of motors in the house and auto.

3. Gas-electric motor generators are used to light airplane beacons which mark major air routes. Discuss how you think these operate.

4. A motor takes more amperage when starting than when running. Suppose a 110 V. motor consumes 2 amps when running constantly. What would it cost to run the motor 100 hours at 3 cents per KWH?

5. Explain how a D.C. toy motor runs.

6. Describe how to oil a motor.

LEARN TO SPELL

generator	synchronous	solar	segments
atomic	ignition	horsepower	sulphuric
commutator	density	cycle	hydrometer

EXAMINATIONS

TEST OF READING ABILITY IN ELECTRICITY

Turn to page 20, Chapter II, section on poles, and read the three paragraphs of the section. Write the answers to the following questions on a piece of paper, numbering the answers the same as the questions. You may refer to the book to help you with your answers.

If you miss more than one or two questions, you should read every lesson several times to be sure you understand the contents. Seek help from your instructor.

1. In what is the magnet to be dipped?
2. Where is magnetism the strongest?
3. What are the ends of a magnet called?
4. Is there magnetism in the earth?
5. What two letters are usually on the ends of a bar magnet?
6. What draws the ends of a magnet toward the north and south poles of the earth?
7. Is the magnetic action the same at both poles of a magnet?
8. Will the N pole of one magnet be attracted by the N pole of another magnet?
9. Is there always magnetism in the earth?
10. Does the earth have polarity?

FINAL COMPREHENSIVE EXAMINATION

Underline the correct answer or complete as indicated. Allow one point for each correct answer to questions not starred. * indicates score allowance of two.

SCORE

1. What type of electricity is lightning? (current) (static) (three-phase) (magnetic) ____
2. Which travels about the same speed as electricity? (sound waves) (thunder) (light waves) (wind) ____
3. Is static electricity used a great deal in modern industry? (yes) (no) ____
*4, 5. Is static electricity the same as current electricity? (yes) (no) ____
6. Which of these men invented the electric light? (Westinghouse) (Bell) (Wright) (Edison) ____

SCORE

*7, 8. Electricity is thought of as flowing out of the (positive) (negative) pole of a dry cell and back into the (positive) (negative) pole ____
9. The current from an auto storage battery is: (A.C.) (D.C.) (three-phase) (110 volts) (23,000 volts) ____
10. The voltage of the modern car storage battery is: (1½ volts) (12 volts) (110 volts) (2,300 volts) ____
11. The amperage of a fully charged car storage battery is usually near: (2 amps) (60 amps) (1,000 amps) (1125) ____
12. Will a 1.2-volt flashlight bulb work efficiently on a 6-volt battery? (yes) (no) ____
13. An hydrometer is used to measure (the amount of charge in a storage battery) (the voltage of magneto) (the capacity of a Leyden jar) (the flow) ____

14. Which of these is used commonly to generate an electric current? (motor) (radio) (generator) (storage battery)
15. A 60-cycle alternating current means that the direction of flow is reversed 120 times per (second) (minute) (hour) (phase)
16. ⌒ is a symbol for a (dry cell) (bell) (outlet box) (crossed wires not connected) (pushbutton)

Underline the correct answer:

17. ___ is a symbol for (crossed wires not connected) (battery) (crossed wires connected) (doorbell)
18. ⊣|⊢ is a symbol for a (buzzer) (ground connection) (battery) (crossed wires connected)
19. ⏚ is a symbol for (crossed wires connected) (battery) (ground) (telegraph key)
20. ⊗ is a symbol for (an outlet) (a dry cell) (a buzzer)
21. Ⓢ is a symbol for a (radio tube) (single-pole snap switch) (doorbell) (ground)
*22, 23. Which carries an electric current the best? (copper) (iron) (tungsten) (nichrome)
24. A conductor is a substance which (allows the current to flow in one direction) (carries an electric current) (stops the flow of a current) (insulates)
25. We measure the diameter of wire with (a wire gauge) (a caliper) (a divider) (a voltmeter)
26. Which of the following will conduct an electric current? (cement) (rubber) (tap water) (leather)

27. Which is the best conductor? (copper) (German silver) (distilled water) (nichrome)
28. Which is the most common size of bell wire? (No. 12) (No. 8) (No. 18) (No. 24)
29. Which is the most common size of house wire? (No. 48) (No. 10) (No. 14) (No. 8)
30. Which wire has the largest diameter (No. 48) (No. 10) (No. 14) (No. 8)
31. Is tinfoil a conductor of electricity? (yes) (no)
32. Three one and a half volt dry cells connected in series will produce how many volts? (1½) (3) (4½) (½)
33. Connect these symbols in parallel to the batteries.

⊗ ⊗ ⊗ ⊗   ⊣|⊢|⊢

34. Connect these in series.

⊗ ⊗ ⊗ ⊗   ⊣|⊢|⊢

*35, 36. Connect these four dry cells to light the 6-volt bulb.

◎ ◎ ◎ ◎   ⊗

37. Give these readings:

10,000   1,000   100   10

May 1 Reading _____

38. Give this reading:

10,000   1,000   100   10

June 1 Reading _____

EXAMINATIONS

SCORE

39. How many KWH were used during May? (_____) _____
40. Amperage refers to (the pressure) or (the amount) of electricity? _____
41. Volts multiplied by amperes equal (ohms) (hours) (KWH) (watts) _____
42. A 200-watt lamp uses how many amperes in a 100-volt circuit? (½) (1) (2) (4) _____
43. A motor using 10 amperes per hour in a 100-volt circuit uses how many watts? (10) (100) (1,000) (10,000) _____
44. 5,000 watts is equal to how many kilowatts? (1) (5) (50) (500) _____
45. If a family uses 142 KWH in a month, what would be their electric bill at a rate of 3½ cents per KWH? ($4.26) ($4.97) ($5.68) ($49.70) _____
*46, 47. Which refers to the pressure of electric current? (amperage) (voltage) (ohms) (electrons) _____
*48, 49. Like poles (attract) (repel) _____
50. The needle of a compass is (an electromagnet) (a solenoid) (a permanent magnet) (loadstone) _____
*51, 52. An electric current flowing through a coil of insulated wire about a core forms (a motor) (an electromagnet) (a compass) (natural magnet) _____
53. The lines of force about a magnet are known as (the positive pole) (the field) (the electrolyte) (galena) _____
54. A compass may be used to determine (the flux) (the polarity) (the size of) (the direction of flow) of an electromagnet _____
55. If the north poles of two bar magnets are brought together they will (attract) (repel) each other _____
56. The earth is a large (electromagnet) (solenoid) (permanent magnet) (Leyden jar) _____

SCORE

57. Is the true magnetic north pole at the same place on the earth as the geographical north pole? (yes) (no) _____
58. A wire under a screw contact goes (clockwise) (counterclockwise) _____
59. Connect "A" to control "B", and "C" to control "D".

*60, 61. Connect so that "S" controls "A" and "B", which contain 6-volt bulbs each. The battery is 6 volts.

*62, 63. Connect so that either "A" or "B" will ring "X" and "Y" together.

64. Is knob-and-cleat wiring permitted in new houses in modern cities? (yes) (no) _____
*65, 66. Using only five wires, connect so that "A" or "B" will turn light "C" on or off independently. (Similar to a stairway light, controlled either upstairs or down. "A" and "B"

are single-pole double-throw knife switches)

67. Do electric motors take more electric power when starting than when running normally? (yes) (no) _____

68. A new car storage battery costs about ($14.00) ($35.00) ($59.00) ($99.00) _____

69. A 50-watt electric light bulb costs about (5 cents) (12 cents) (35 cents) (75 cents) _____

70. A new ¼-horsepower, 110-volt A.C. motor costs around ($1.00) ($10) ($45) ($100) _____

71. A good electric gun-type soldering iron costs about ($8.00) ($20.00) ($30.00) ($40.00) _____

72. A new extension cord to be used for a floor lamp complete with plug, 8 feet of wire, and socket can be bought for (from 20 cents to $1.00) (about $4.00) (from $4.00 to $5.00) (over $5.00) _____

73. When a fuse blows, first: (find the cause and correct it) (call an electrician) (replace with a new fuse with lower amperage rating) (replace with a fuse with higher amperage rating) _____

74. Flush switches are more often used to (make and break a circuit) (bend pipe conduit) (fill a storage battery) (shunt the condenser) _____

75. Alternating current may be stepped up or down with a (dynamo) (fuse block) (transformer) (derrick) _____

76. Resistance wire is commonly used in (the electric motor) (the toaster) (the electric sweeper) (electric clock) _____

77. Poor connection may cause (an arc to start a fire) (spontaneous combustion) (a flux) (a primary flow) _____

78. Joints are soldered because (they look well) (they tape better) (they are safer and stronger) (cost less) _____

*79, 80. What size fuse would you select for your home in which the total amperage of all appliances equalled 12 amps? (10 amps) (20 amps) (40 amps) _____

81. An interrupted current of 6 volts flows in the primary circuit of an induction coil of 100 turns. The secondary coil has 1,000 turns. Theoretically, the output voltage is (.6) (6) (60) (6,000) (1.6) (_____) _____

82. In the automobile the secondary current from the ignition coil flows first to the distributor and then to the (spark plugs) (fuse) (battery) (circuit breaker) before being grounded _____

83. An A.C. clock uses for power (static electricity) (alternating current) (direct current) (three phase) _____

84. Are there any electric motors used in the modern automobile for any purpose? (yes) (no) _____

85. Sound waves are changed to electric impulses in a telephone transmitter by the (fuse plug) (photo-electric cell) (varied pressure on the granules) (use of neon) _____

86. Do doctors use electrical machines to promote health or cure diseases? (yes) (no) _____

87. Does an electric clock take as much power as a radio? (yes) (no) _____

88. Can a telegraph sounder be used in place of a relay? (yes) (no) _____

EXAMINATIONS 159

SCORE

*89, 90. In the telephone are the carbon granules located in the receiver? (yes) (no) _____
91. The kilocycle is (100 cycles) (.001 of a cycle) (1,000 cycles) (10,000 cycles) _____
92. The electrical waves sent from distant broadcast stations and received by home radios are increased or amplified by the (rheostat) (earphones) (vacuum tubes) (condenser) _____
93. Most American radio stations broadcast on wave lengths

SCORE

which are measured in (feet) (miles) (inches) (meters) _____
94. The worker in the electrical profession requiring a high school and university education is called a (linesman) (electrical engineer) (motor repairman) (ignition mechanic) _____
95. The worker in the electrical trades requiring a trade school or apprentice education is called (an electrical engineer) (an electrician) (civil engineer) (designing engineer) _____

INDEX

Alarm switches, 63
Alternating current, 30
Ammeter,
  connected in circuit, 34
  purpose of, 137
  symbol for, 35
Amperage, 29, 34
Amplifier
  one tube, one transistor, 145
Appliances, electrical, 70, 71, 73
Arcing, 50
Atom, 48, 144
Atomic power, 129, 141, 150
Automobile electricity, 136

Battery,
  storage, 138, 139
  symbol for, 35
Bells, 53
Bell-wiring problems, 60
Blinker lights used for a code, 114

Cells,
  dry, 31
  symbol for, 35
  using, 55
Chimes, electrical, 42
Circuit,
  ignition, 132
  lighting, 133
  parallel, 57
  series, 58
  starting, 134

Code,
  Continental, 116
  Morse, 116
  National Underwriters, 50
Coils,
  field, 127
  primary, 41
  secondary, 41
  spark, 44
Color TV, 126
Compass, 19, 27
Conductors, 47
Conduit, 105-109
Connectors, 95, 105
Cords,
  extension, 86
  flexible heating-appliance, 72
Core, magnetic, 37
Current, types of, 30
Cycle, 30

Density, 135
Direct current, 30

Electricity,
  current, 29
  in cells, 31
  static, 28
  store of, 11
Electromagnet, 37, 40
Electronics, 141
Electrons, 11, 31, 48, 78
Electron theory, 48

INDEX

Field coils, 127
Field of force, 22
Flashlight, 79
Fluorescent lamps, 89
Fuses, 74, 137

Galena crystal, 122
Galvanometer, 31
Generators, 30, 130
Glow plug engine, 140, 141

Horsepower, 126
Hydrometer, 139

Induction coil, 41, 44
Insulation materials, insulators, 47, 49

Keeper, to retain magnetism in magnets, 20
Kilowatt, 84

Lasers, 151-153
Lines of force, 37
Loadstone, 19

Magnets, 19, 20, 26
Magnetic field and flux, 23
Magnetism, 19-25
Magnetite, 19
Meter reading, 34, 36, 84
Motors,
  synchronous, 134
  types of, 130, 131

National Underwriters, 50

Occupations, professions and trades, 15, 16
Ohms Law, resistance, 66, 69

Parallel,
  circuits, 57
  lights in, 81
  wiring of cells, 32, 33
Polarity, 21, 25, 39
Poles, 20, 23
Power, electrical, 129
Proton, 48

Radar, 121
Radio, 122, 123, 126
Railroads and electricity, 141
Ready-cut materials, 10
Relay telegraph, 114
Resistance, 66
Rheostat, 68
Rocket power, 150

Series,
  circuits, 58
  lights in, 81
  wiring of cells, 32
Shock, nature of, 43
Signal alarms, 63
Sockets, lamp, 87
Solar power, 150, 151
Soldering, 95
Solenoid, 40
Spark plug, 133
Splicing,
  rat-tail, 93, 94
  tap, 51, 94
  Western Union, 51
Storage battery, 138, 139
Switches,
  automatic, 120
  entrance, 103, 105
  feed-through, 88
  for house wiring, 99
  mercury, 101
  thermostatic, 102
  three-way, two-way for light, 80, 100

Tape recorder, 127
Telegraphy, 113-116
Telephone, 118, 122
Teletype, 124
Television, 125, 126
Terminals, 55
Thermostat, 104
Tools for electrical shop, 9
Toy motors, 134
Transformer, 44, 55, 105
Transistors, 144, 149

Underwriter's knot, 87

Vacuum tube, 142
Voltage, 29, 34, 68, 104
Voltmeter, 34, 35

Watt and kilowatt, 82, 84
Welding, arc, 75
Wire
  gauge, 49
  house, 92
  nichrome, 68
  table, 67
Wiring,
  conduit, 105
  knob-and-cleat, 98
  knob-and-tube, 97
  open, 92
  with conduit, 109